Praise for *American Hemp Farmer*

"A fantastic piece of Americana that shows the way to a sustainable future."
— **DAVID BRONNER**, CEO, Dr. Bronner's Magic Soaps

"With *American Hemp Farmer*, Doug Fine shows he is not just our preeminent hemp author, he is one of the most important authors of our time. As I've watched him leap between tending goats on his Funky Butte Ranch and hemp fields in Hawaii, Oregon, Vermont, and who knows where else, it sometimes occurs to me that he might be the most interesting man alive. The resulting book is an absolute must-read."
— **ERIC STEENSTRA**, executive director, Vote Hemp

"After 83 years of prohibition, cannabis's emergence from the underground has sparked a gold rush that has every farmer, wannabe farmer, and agricultural entrepreneur rushing to stake their claim. With *American Hemp Farmer*, Doug Fine makes an incredibly well-written case for a regenerative agriculture–based, small- to mid-scale approach to the industry that prioritizes quality of over quantity, and where soil carbon sequestration is a bottom-line goal. Humorous, timely, and important."
— **JEFF CARPENTER**, coauthor of
The Organic Medicinal Herb Farmer

"*American Hemp Farmer* would have been in George Washington's library. President Washington grew hemp and was a passionate, regenerative agriculturist. Washington sought advice from those that practiced their trade. Doug Fine's *American Hemp Farmer* is a scholarly, practical, and impeccably enjoyable work and a must-read for those who cultivate hemp or are interested in leaping in."
— **DEAN NORTON**, director of horticulture, Mount Vernon Estate

"In his latest, author Doug Fine—a modern day Johnny Hempseed—has painstakingly penned a love letter to the cannabis plant and all those who tend it. Doug details the beneficial and no longer forbidden relationship between cannabis and humanity and how together there is a path to rejuvenate the entire planet. As a state hemp program administrator, I hope every hemp farmer and policymaker reads this book carefully. It details a roadmap for success, for farmers and the planet. And that's probably because Doug doesn't just write about hemp, he lives it."

— **CARY GIGUERE**, hemp program coordinator,
Vermont Agency of Agriculture

"As a hempcrete homeowner, I'm proud to keep *American Hemp Farmer* on my shelf as *the* must-read book on hemp. Someday we may even see NBA arenas built from hemp. But for now, Doug should be prepared to lose more money at the poker table that sits on the hemp floor of my hemp-paneled card room."

— **DON NELSON**, two-time
NBA Hall of Fame inductee

AMERICAN
HEMP
FARMER

Also by Doug Fine

Not Really an Alaskan Mountain Man

Farewell, My Subaru:
An Epic Adventure in Local Living

Too High to Fail:
Cannabis and the New Green
Economic Revolution

Hemp Bound:
Dispatches from the Front Lines
of the Next Agricultural Revolution

AMERICAN HEMP FARMER

ADVENTURES *and* MISADVENTURES *in the* CANNABIS TRADE

DOUG FINE

CHELSEA GREEN PUBLISHING

White River Junction, Vermont
London, UK

"Cannabis, *Cannabis sativa*" by Ida Pemberton, reprinted with permission
by University of Colorado Museum of Natural History Herbarium (COLO).

Project Manager: Alexander Bullett
Acquisitions Editor: Brianne Goodspeed
Editor: Michael Metivier
Copy Editor: Eliani Torres
Proofreader: Deborah Heimann
Designer: Melissa Jacobson
Indexer: Shana Milkie
Page Composition: Abrah Griggs

Printed in Canada.
First printing April 2020.
10 9 8 7 6 5 4 3 2 1   20 21 22 23

Our Commitment to Green Publishing
Chelsea Green sees publishing as a tool for cultural change and ecological stewardship. We strive to align
our book manufacturing practices with our editorial mission and to reduce the impact of our business
enterprise in the environment. We print our books and catalogs on chlorine-free recycled paper, using
vegetable-based inks whenever possible. This book may cost slightly more because it was printed on paper
that contains recycled fiber, and we hope you'll agree that it's worth it. *American Hemp Farmer* was printed
on paper supplied by Marquis that is made of recycled materials and other controlled sources.

Library of Congress Cataloging-in-Publication Data
Names: Fine, Doug, author.
Title: American hemp farmer : adventures and misadventures in the cannabis trade / Doug Fine.
Description: White River Junction, Vermont : Chelsea Green Publishing, [2020]. | Includes
   bibliographical references and index.
Identifiers: LCCN 2019058309 (print) | LCCN 2019058310 (ebook) | ISBN 9781603589192
   (paperback) | ISBN 9781603589208 (ebook)
Subjects: LCSH: Hemp—United States. | Hemp farmers—United States. | Hemp industry—
   United States.
Classification: LCC SB255 .F55 2020 (print) | LCC SB255 (ebook) | DDC 633.5/30973—dc23
LC record available at https://lccn.loc.gov/2019058309
LC ebook record available at https://lccn.loc.gov/2019058310

Chelsea Green Publishing
85 North Main Street, Suite 120
White River Junction, VT 05001

Somerset House
London, UK

www.chelseagreen.com

For the soil and the soil builders.
Thanks for doing your job well.

For the goats and the bear lost in the 2013 Silver
wildfire. We're trying to address the causes herein.
Hence the appreciation for the soil builders.

And for all folks, rural or urban, who decide to give
farming one more shot before caving to un-developers.
This time, the farmers are in charge.

The farm holds the key to human happiness.

—WILLIAM J. HALE, director of organic chemistry research,
Dow Chemical Company, in *The Farm Chemurgic:
Farmward the Star of Destiny Lights Our Way* (1934)

It's beginning to look like it might work.

—JOHN HICKENLOOPER, former governor of Colorado,
who opposed his state's cannabis legalization in 2012,
on the $100 million in annual tax revenue
the industry was generating by 2016

CONTENTS

Refugee Bear

Six years ago, a bear fleeing a wildfire in our New Mexico back-yard killed nearly all of my family's goats in front of our eyes. It wasn't the bear's fault; he was a climate refugee. It was June of 2013, and drought had weakened the ponderosa pines and Douglas fir surrounding our remote Funky Butte Ranch. Beetles took advantage, and all of southern New Mexico was a tinderbox. Ho hum, just another climate event that until recently would have been called a "millennial" fire.

The blaze cut a 130,000-acre swath that year, poisoning the air before the monsoon finally arrived about half a day before we would've had to evacuate. But it was too late for the large juvenile black bear, who'd lost his home and his mind. He didn't even really eat most of the goats. We lost all but one of the animals that provided our milk, yogurt, and ice cream.

Baby Taylor Swift survived, but Bette Midler, Stevie Nicks, and Natalie Merchant (who loved meditating with me of a morning) perished, as did the bear several weeks later, care of a Game & Fish marksman, upon going after a dozen of our neighbor's sheep.[1] Ever since, my sweetheart and I have had to keep a constant eye on our human and goat kids. We react like a frenzied SWAT team to any unusual noise up in the epony-mous buttes above our small adobe ranch house. We've had our climate change Pearl Harbor—the event that shifted us into a single-minded new normal. If you haven't had yours yet, you probably soon will.

This is the paramount reason I'm an overworked employee of the hemp plant: The people I care about most are one blaze away from join-ing the world's 20 million climate refugees. At least I get the pleasure of putting "goat sitter" under occupation on my tax form.

There's nothing like wildfire-fleeing bears attacking your livestock before breakfast to hammer home the fact that humanity is in the bottom of the ninth inning with two outs. The conflagration convinced me that I had to do something, personally, to work on this climate change problem. After some research about carbon sequestration through soil building, it became clear that planting as much hemp as possible was the best way to actively mitigate climate change and help restore normal rainfall cycles to our ecosystem.

At least the fire's timing was good. Hemp was de facto legalized for "research purposes" in 2014, two months before the publication of my earlier book about hemp, *Hemp Bound*. I've spent the five ensuing years not just covering the new industry but joining it: developing genetics in Oregon and a farm-to-table product in Vermont; consulting, filming, and speaking all over the world; working on university research in Hawaii; and teaching a college course.

But planting hemp and making a living at it can be two different endeavors. This book blueprints possibilities for independent farmers like myself who'd like to do both, particularly on their own land. If a lot of things go right, an independent farmer (or a farmer cooperative) can make a viable living on a small number of acres. That ain't exactly the way agriculture has been going for the past century. Just how many acres depends most of all on the part or parts of the cannabis plant you are cultivating (seed, flower, fiber, root). Another variable is whether you're planning to create a value-added product. A third is if you're going at it alone or in partnership with others.

Hemp markets are diverse enough that I've met farmers who have developed a viable business plan for a 1-acre harvest at the same time that there are independent farms in Oregon, Kentucky, Montana, and Colorado cultivating in the 2,000-acre realm. We'll focus on a 20-acre enterprise throughout our adventures in these pages, as we glide from soil prep through cultivation and on to strategies for marketing final products. In many of the season's phases, the discussion will be scalable to either larger or smaller enterprises.

Though I still consider myself a hemp journeyman, I've got a dozen crops under my belt, across varied soils, climates, and laws. So we'll explore the most illustrative ways that this plant has put me and others through the wringer during each part of the season. Our planting saga will come from one year's cluster of unasked-for lessons (tractor issues in Oregon), our harvest from another year's adventure half a continent away (a snowstorm hurrying us in Vermont). Then we'll follow the efforts of several pioneer hemp-farming enterprises to bring the resulting farm-to-table products to the world. When we emerge we'll have survived an entire season.

For those who don't want to make a living with hemp work but would like to support the farmers who do or perhaps grow their own ancient superfood while sequestering some carbon, the lessons from my ongoing immersion are the same. Plus, for backyard gardeners and pros, working in a hemp field is the most fun you can have outside the bedroom.

Even as I relate the experiences of a half decade in hemp, this book also reflects life unfolding in real time. That's because when you're strapped in for the roller-coaster ride of a major industry's first wild years, new realities arise almost daily on all fronts. In the case of hemp, cultivation lessons, permitting and marketing rules, and promising markets are all in constant flux. Perhaps most important, hemp was just fully legalized for commercial purposes in the United States a few hours before the 2018 winter solstice, the day I started working on this project.

Thanks to a little 28-page provision tucked into the 807-page, $867 billion Agriculture Improvement Act (2018 Farm Bill)—which became law while I was extracting our newest Houdini of a goat kid, Julie Andrews, from the ranch's winter cover crop—hemp's federal oversight has been transferred back from the purview of the Justice Department to that of the Department of Agriculture (USDA). This is where it belongs—hemp being just another farm product.

For three-quarters of a century, cultivating hemp (today meaning nonpsychoactive varieties of cannabis) had been functionally illegal in the United States. This started in large part because of a bureaucratic budget shuffle. The guy who ran the federal alcohol prohibition program during its final stages, Harry Anslinger, needed a job for himself and existing staff, so he and some friends in the yellower media set about inventing a problem with one of humanity's longest-utilized plants.

Under the 2018 Farm Bill provision, our public servants at agencies like the Food and Drug Administration (FDA) will chime in on edible products. In fact, FDA honchos were already issuing menacing memos about being the new sheriff in town, just minutes after law enforcement agencies such as the Drug Enforcement Administration (DEA) had been freed to focus on the opioid epidemic and other real problems.

Those administrative shuffles mean that, in these pages, we'll have to spend a portion of our hemp year off the field, learning how to deal with—and shape—all kinds of regulations: farming regulations, nutritional-supplement regulations, hemp-testing rules. Another way of putting this is that the entrepreneurs and activists who worked for decades to bring about this momentous legalization—and who were justifiably blowing up my phone with a barrage of emoji-laden "Victory!" notes on that joyous day the 2018 Farm Bill passed—are about to have a "be careful what you wish for" adjustment. But that's okay, and to be expected. Collectively we independent farmer-entrepreneurs (and the customer base that supports us) will make sure the emerging industry rules work for our farm-to-table craft sector. That way we can rebuild both soil and rural communities.

As I type here on the ranch, 10 months after that legalization solstice, the unusually orange orb of a near-full moon is rising outside my office window as though in celebration—one more crop has come in. The long nightmare of cannabis prohibition is over. Its three-generation duration is to our advantage: We can shape this industry any way we like.

—DJF
Funky Butte Ranch, New Mexico
October 2019

CHAPTER ONE

Be First, Better, or Different

New Mexico, 2019

In the week between learning about the discovery and yelling about it in San Francisco, he'd bought all the picks and shovels in the city.

—PETER YANG, writing about store owner
Sam Brannan's actions in the wake of
the 1848 California gold strikes[1]

If we attempt to pinpoint why hemp is about to become the fastest agricultural industry ever to reach a billion dollars in annual sales, it might be because humans have an embedded genetic memory about the plant.[2] Hemp has been a "camp follower" crop, say anthropologists, since before the arguable misstep of sedentary agriculture.[3] Michael Pollan argues in *The Botany of Desire* that we actually co-evolved with cannabis and other plants. To him it just makes Darwinian sense—if a plant wants these apes to keep carrying and planting it around the planet, that plant will do its best to give them things they want: roofs, sandals, superfood, party favors. Not bad from one seed.

In an era when any material is a click away, I utilize hemp every day, strictly for performance reasons—it beats the competition: might be the plant's seed, fiber, or flower (see the image on page 8). Often all three. I eat it, wear it—I'm about to patch my porch with a homegrown hemp fiber plaster. My laptop case is made of hemp fiber, too—I like to think hemp's microbial-balancing properties protect me from disgusting airplane tray tables.

There is physiological evidence to support Pollan's co-evolutionary theory. In 1992, the Israeli researcher Raphael Mechoulam discovered that we are all born with receptors for the compounds in cannabis flowers and some other plants (collectively called cannabinoids). Our built-in cannabinoid receptors constitute what is known as our *endocannabinoid system*. Put simply, these receptors prepare our bodies to receive the properties of cannabinoids. You can think of them as Velcro, waiting for, say, the pain-relieving components of cannabis to be introduced when we bark our shin. All mammals, amphibians, reptiles, birds, and fish have endocannabinoid systems. There's preliminary research into whether some invertebrates also have them.[4] If they do, that sure dates our relationship with the cannabis plant way back in history.

Modern farmers reveal their encoded love for hemp without always realizing why. When my Kentucky colleague Josh Hendrix took me to an antebellum barn that sported a World War II–era hemp brake used to prepare rope for navy rigging, he demonstrated the recently rediscovered, calf-sized wooden device as though he had been using it for years.

"Hemp is deeply rooted in rural Kentucky culture," he told me. "Most people would call this a tobacco barn. But before that it was a hemp barn."

Seventy-seven years of cannabis prohibition, in other words, are a blip in humanity's eight-millennia relationship with this plant.

From an economic standpoint, said Steve DeAngelo, a prominent cannabis activist since the 1970s, "the cannabis industry's sustained double-digit growth curve is almost unprecedented in modern business history, and that's before the whole plant is legalized federally." The phenomenon he describes is not limited to North America—a Moroccan farmer named Adebibe Abdellatif flew halfway across the globe on his own dime to attend a 2017 United Nations cannabis session, where he told me his motivation was to ensure that the global hemp reemergence "is steered from the farm."

DeAngelo cofounded the Harborside nonprofit dispensaries (originally Harborside Health Center) in California in 2006, and is in a unique position to characterize the industry's growth curve. "Because of its breadth of applications, cannabis/hemp is the most disruptive economic development since Silicon Valley edged out blue chips," he told me. "To say we're in our infancy is an understatement when it comes to this plant's uses and markets."

The biggest driver of hemp since (and only since) the first research-only Farm Bill has been cannabidiol, popularly known as CBD. The market for this valuable, hemp-flower-derived nutritive supplement and topical application is growing 23 percent annually, and is on its way to being firmly established in the healthy person's wellness lexicon, the way that *omegas* and *aloe* are.

At the moment, it's not an exaggeration to call the CBD market a gold rush. It is one compound in the family of cannabinoids that resides in the female hemp flower. As Arthur Rouse, a Kentucky journalist who has been documenting the fits and starts of hemp's reemergence since the 1990s, sees the current reality on the ground: "A few veins have been struck [in the hemp flower]. Now everyone's flocking to the site of the first strike: CBD."

CBD is terrific. It's a compound that is genuinely benefiting millions of people. My own cannabinoid intake serves as a dietary supplement, part of my health maintenance program. CBD itself is not temporary; what is temporary is CBD-only mania and, for farmers, high wholesale prices.

Previous gold rushes, such as California's in the 19th century, provide clear lessons. Gold, of course, was and is still being traded long after most '49ers went bust—it wasn't the prospectors who benefited from commodities markets. Only a few made a strike in California, and half of them got hoodwinked out of it by shady middlemen. We're going to avoid that outcome with regard to hemp. Long-term, maximum farmer benefit is our goal for this economic boom.

The types of people who got reliably rich off the 19th-century North American gold rushes were the same ones who get rich off any boom: middlemen (crooked or legit), real estate developers, and the folks selling the shovels, pickaxes, tents, pack mules, and sacks of flour and coffee. Today hemp has its own middlemen, real estate developers, and shovel sellers, but they're called extraction salespeople, CBD wholesalers, warehouse lessors, and venture capitalists.

Some of these folks are honest and well meaning. But there's no denying that elements of the hemp renaissance have all the makings of

From left to right: male cannabis flower; male plant (with depiction of bast and hurd fiber); seed; seed cross-section; female plant; female flower; trichome-laden female flower prior to pollination. *Illustration by Ida Pemberton, reprinted with permission by University of Colorado Museum of Natural History Herbarium (COLO).*

The Architecture of the Digital Age's Hottest Commodity

Seed: A superfood with perfect omega balance and loaded with minerals. Although fewer than 1 percent of U.S. homes today stock a hemp food product, those that do sure use a lot of it; edible hempseed, hemp protein meal, and hempseed oil already constitute a billion-dollar market in North America, one growing by double digits annually. In the near future (as in most of the past), hempseed derivatives will also provide a regenerative source for industrial solvents, resins, and glues.

Flower: Source of the current CBD gold rush, the crystalline bulbous trichomes that line hemp's female flower contain more than 111 known compounds, called cannabinoids, many with beneficial properties, and many of which will feed future gold rushes. Absent from the hemp market just five years ago, hemp-flower products now represent 80 percent of the fast-growing industry. Overnight, the majority of hemp cultivators have migrated from the 8,000-year-old dioecious (male and female) mode of hemp cultivation to sinsemilla (female only, literally "without seed") cultivation. That's because they are interested only in the CBD gold rush. For now.

Stalk (Fiber): Feedstock for tomorrow's cars, space modules, and batteries, and for today's high-end homes and horse bedding. Hemp fiber, alongside other biomaterials, will be a key source of humanity's migration from fossil fuels and petrochemicals. To be viable, fiber applications require large-acreage cultivation. Anything petro-plastic can do, hemp fiber and other biomaterials can do better. The hemp stalk contains two distinct components: the long strips of strong outer bast fiber, and the remaining inner core, called hurd or shiv. Each has different properties serving distinct industrial needs.

Root: Saving the planet by sequestering carbon (three billion tons annually when worldwide topsoil is rebuilt by just one inch). Hemp's unusually long taproots help create the belowground climate to allow the world's struggling soil to rebuild.

one of those bursts of irrational exuberance that accompany any market bubble. The sad reality is that many of the early hemp players one sees sponsoring trade show lanyards in 2019, inexorably churning through angel investment and gunning for CBD dominance, aren't going to be with us by 2025. The proverbial wheat will separate from the chaff (or in the case of the plant we are discussing, the bast will separate from the hurd, though both of these are valuable).

Even though it will require an industry that markets all parts of the hemp plant to sustain a new farming economy and sequester enough carbon to ensure a habitable planet, let's start with the flower and discuss the market for the cannabinoids therein, because CBD is one of them. And CBD is about the hottest business and health story in the world today. Houston, Rome, Santiago, Tokyo, and Cape Town all have CBD stores. The World Health Organization has declared CBD safe. Mike Tyson, who has his own cannabis line, calls it "a miracle" for someone who's had his cranium rattled a few times.

Cannabinoids comprise a growing group of 111 known compounds found in cannabis, other plants (such as cacao, pepper, and echinacea), and interestingly, endogenously in mammalian mother's milk. THC (tetrahydrocannabinol, the "psychoactive" component in cannabis) and CBD are the best known of these cannabinoids.[5] My own favorite cannabinoid at the moment is CBC (cannabichromene), a nonpsychoactive compound showing analgesic properties, as well as anecdotal evidence of anti-inflammatory and muscle-relaxing effects.[6]

Plants including hemp produce cannabinoids because they serve a range of purposes including predator defense, climactic adaptation, and pollinator attraction.[7] And also, as Pollan postulates, to please us.

Cannabis/hemp flowers even smell appealing. So appealing that I routinely have to check myself in the field, lest I eat the profits. The flowers contain terpenes (*terps* to those in the business), fragrant hydrocarbon-based compounds that are found in the essential oils of many plants. They might enhance the properties of other components of a plant (much academic research on this subject is still in progress), but their smells alone add to the value of a cannabis flower. Some farmers already breed just for terps. You can buy terpene-laden cannabis in dispensaries. Their scents and properties vary widely. (My favorite is one called pinene.)

If CBD is the mine where most prospectors, both independent and would-be giants, are currently staking their claim, it's a near certainty that this won't be the case in five years. Change being the only constant, I feel safe declaring that a previously unimagined market sector will emerge by 2025. I hope it's recyclable, next-generation hempen battery components, a hempseed diet craze, or a bunch of next-wave cannabinoid-terpene combinations.

When cannabinoids and terpenes work in concert, it's known as the *entourage effect*, a key argument for thinking beyond one cannabinoid when it comes to hemp product efficacy. I know I wonder about the properties embedded in blended cannabinoids—say, #7, #42 and #81—grown in a high-pinene flower. Efficacy might reside not necessarily in the sheer number of milligrams of CBD in your tincture but in the interplay of many cannabinoids in ideal ratios.[8]

This next-phase industry morphing we're about to see will favor flexible, independent farmer-entrepreneurs. As my Alaskan river guide instructor taught me on the very first day of training back in 2004, "Learn to look three turns ahead."

Folks looking for a quickstart guide to capitalizing on the CBD craze? That is not three turns ahead. That's the momentary straightaway—you might well crash into the bank before the first turn. Especially if you're relying on the temporarily inflated wholesale market.

But even if you've come to this book looking for the Powerball numbers required for a CBD jackpot, I hope you'll approach these pages with an open mind, ultimately absorbing the following message very carefully: Yes, the CBD market is predicted to grow to $1.65 billion by 2021 from $291 million in 2017.[9] But, as with previous gold rushes, independent farmers (the prospectors) won't be earning most of it, unless we market our own products regionally, rather than wholesale our harvests to glean whatever living far-off commodities markets dictate.

For each of the past five years, hemp acreage in the United States has more than doubled, a trend likely to continue for another half decade at least. But that means something only if the industry sets its baseline standards according to regenerative principles. Fortunately for humanity, hemp's return coincides with (and informs) the reawakening of a global awareness that the Earth is a system like a store's shelves. Barring space mining or our evolution into some kind of pure astral awareness that obviates the body's needs, our planet's continually renewed resources are the only possible source of re-stocking everything that keeps the species surviving and thriving.

Vermont, one of the states where I cultivate, has focused its hemp program policy on independent, small-acreage farmers since before the 2014 federal Farm Bill provision. The state's hemp administrator, Cary Giguere, is on message in his awareness that our best strategy for farmers, climate, and the long-term economy likely resides in a "biomaterials economy," one based on regeneratively grown plants, algae, and other God-given supplies.

"The monoculture era hasn't been working out for farmers or the planet," Giguere said. "Synthetic pesticides and herbicides tend to only work for a while."

He's right. A 2019 United Nations report found that 22 percent of the 2.7°F temperature increase the planet has experienced in the past century and a half is due to outdated agriculture and forestry practices.[10]

The term *regenerative agriculture* was coined by Bob Rodale in the 1980s, as part of his "beyond sustainable" farming theory. Regenerative agriculture was necessitated, Rodale felt, by the small and declining amount of worldwide topsoil remaining at the end of the last century.[11] Today, the term *regenerative* is both widely used and malleable enough that folks often ask me, "Do you mean sustainable / organic / recyclable / compostable / fair trade?" when I pepper a talk with the word. To which I answer, "If, in the course of your everyday business processes, what you're doing will be good for humanity's well-being generations down the line, it's regenerative."

In my own hemp enterprise, regenerative practice means trying to be aware of my impact in everything I do. From cultivation to packaging to delivery, it means rebuilding as I produce, so I can produce again. It includes practices like reduced–fossil fuel farming and compostable packaging.

Regenerative entrepreneurs are this book's protagonists. They already populate a substantial hemp-industry niche. Independent farming might even be the largest component, one with a real chance to be at once the most lucrative industry sector and the one most essential for the survival of our species.

The survival aspect is fairly easy to quantify. A growing body of research suggests that each cubic inch of topsoil we restore of the world's farmland sequesters up to three billion tons of carbon annually.[12] And hemp's substantial taproots are absolutely stunning at creating the conditions that allow for the building of topsoil. We're all wise to root for an industry that helps with climate stabilization. If the regenerative farming mode catches on, farmers might even sequester sufficient carbon to buy us humans a crucial century to get our underlying infrastructural cards in order—the goal being to thrive, rather than panic, as we glide into the post-petroleum future.

Living, as we do, in the era in which *Merriam-Webster's Collegiate Dictionary* added the term *bug-out bag*, there's no longer time for operational hypocrisy and greenwashing. For "We'll import offshore CBD for a few years until we can afford to support local organic farmers." Or "We'll make our packaging compostable when we have some money in the bank." We're all one fire or flood from having to bug-out. Solutions to the climate crisis have to begin with the birth of every business. No enterprise I've encountered is perfect, and we don't need to beat ourselves up if we find ourselves plugging gaps as we go. But a fundamental commitment to running completely regenerative operations must begin at launch.

The "lucrative" side is where the necessary win-win of regenerative entrepreneurialism resides: Independent hemp farmers are already showing that small-acreage, farm-to-table products are nearly always superior to mass-produced ones, the way fresh-squeezed orange juice beats frozen concentrate. Without that marketplace superiority, merely saving humanity would be a tough sell to folks entering the industry as economically stressed family farmers. The essential point is that regenerative values can still be entrepreneurial. Everyone wants to make a living.

Also important to keep in mind is that hemp is merely leading the way in this wider migration back to biomaterials as our primary industrial feedstock. This decade's two Farm Bill provisions have released the first arrow of the coming regenerative-biomaterials-era barrage. Soon, if we're successful in our execution and messaging, the processes hemp's pioneers are developing will seed the industrial pipeline in areas well beyond one plant. And not just farming processes but also enterprise structural processes (like profit sharing and the values embedded in B corporations and co-ops) and financial services processes (bye-bye, crappy banks).

So thank a prohibitionist: By keeping this plant out of legal markets for three-quarters of a century, he's handed us the opportunity to launch without the "but we've always done it this way" ball and chain. At the same time, the unleashed hemp industry is expanding and evolving so rapidly that there almost certainly will be a next hot app or three in play by the time this book comes out.

Relying on wholesale CBD is not a viable game plan for most independent farmers for reasons beyond even the coming fungible market price correction. In 25 years, CBD itself will be regarded the way the transistor is in the tech sphere today: very useful, a key building block in the early stages of the modern industry, but such a small part of the evolving picture as to be almost quaint, like the early video game *Pong*. So save a couple of your early bottles of expensive CBD; they'll be valuable collector's items one day. Now is the time to sidestep the CBD-only herd and explore the countless other opportunities that the hemp plant provides. Heck, CBD represents less than 1 percent of known cannabinoids. And the flower is just one of the four useful parts of the cannabis plant's architecture (alongside seed, fiber, and root).

Flower entrepreneurs weren't even invited to most hemp industry trade group conventions until 2014. Now CBD (and such ancillary products as extraction equipment) represents as much as 80 percent of the industry, and three out of five booths bought at industry trade shows, according to Lizzy Knight, cofounder of the NoCo Hemp Expo. Given all that hemp has to offer, that's not a rational leap. That's a gold rush leap. That's a bubble.

To look at it from another angle, from 8,000 years ago through 10 years ago, male plants (or male parts of hermaphrodite plants) grew in 100 percent of hemp fields. Today they grow in 20 percent.

As Mark Reinders, managing director of Europe's oldest hemp company, HempFlax, reminded me half a decade ago when I started researching hemp, "Success in the early modern hemp industry comes to those who are constantly ready to pivot."

That advice is probably a truism in any new industry, especially in the digital age. When I interviewed Reinders at the HempFlax warehouses

"To Anybody Thinking About Hemp"

On September 4, 2018, I got a voice mail from 84-year-old Wendell Berry, author of *The Unsettling of America* and many other books. Berry, perhaps our greatest living farmer-philosopher-poet, was calling me because I had written him (on hemp paper, of course) to invite him to speak at a hemp conference I was helping organize near his Kentucky home.

Mr. Berry's message, related in his oscillating octogenarian timbre, is the primary theme of this book: Value-added marketing and control of production and distribution by farmers are crucial to our success. I've saved the voice mail and here's a transcription of the meat of it:

> I would like to say to anybody thinking about hemp, that if everybody grows it [to sell to middlemen and wholesalers], it will eventually drive the price down, and you'll be in the same fix as the soybean people. So you need to be thinking about production control, the way that Organic Valley has thought about it [by marketing its own products], as an outgrowth of the old tobacco [cooperative] program.

In the 1930s, Berry's father and brother helped establish the Burley Tobacco Growers Cooperative, aimed at circumventing the exploitive middlemen who were keeping farmers poor by controlling prices. Also called the Producers Program, the co-op's existence overlapped almost completely with cannabis's 77-year prohibition. As tobacco fades, the model is ready to be scooped up by hemp farmers, worldwide. In fact, there are already hemp co-ops active in Kentucky and Colorado.

Berry's message is not just for hemp farmers. It resonates in my own remote ranching valley. When interviewing him for a National Public Radio (NPR) story about declining water supplies in the American Southwest, I noticed that my neighbor, Dennis Chavez, an old-timer, had an entire orchard of gorgeous, nearly purple heirloom apples that date back to Spanish varieties. While I crunched into one, he told me about the day, in 1976, when the regional supermarket buyers told him that all commercial apples in New Mexico would henceforth be coming from California. What a loss, in taste alone, I thought.

"We had a fine little industry going here," Chavez told me. "It disappeared overnight. That's when you realize that the farmers aren't in charge of their livelihoods."

We're working on that, by listening to Wendell Berry and a couple of other prophets. This time, Dennis, the farmers are in charge.

in Holland in 2013, his mechanics showered sparks on us as they frantically retrofitted the company's harvesting equipment in order to capture this strange new part of the plant, the flower.[13]

So what is the wise move, if not churning out flower for the CBD wholesale market? For the answer, we turn to the great American artist and entrepreneur Dolly Parton. When I was a kid, I once heard her tell an interviewer something that has always stuck with me.

"Honey," she said, hips a-shakin', finger a-waggin', "if you want to succeed, you've either got to be first, better, or different."

Create your own specialty brand, in other words. Personally, I'm aiming for "better," with a little bit of "different": By infusing the unusual flower I grow in the hempseed oil pressed from the same crop and doing it in small batches, I think I've created a distinct product deserving of a bit of shelf space.

As have many others. Yes, this book presents the thesis that the independent craft sector is already hemp's leading brand. But the fact is, none of us is the first into CBD. If you're a small-acreage hemp farmer, someone else is going to supply Walgreens and the inevitable Coke CBD. What you can be is part of your region's Organic Valley, Ben & Jerry's, or Burt's Bees. In a world of McCrap options for most things, more and more people crave the real thing.

Even in my product bottling, I don't let the customer forget that message for a second. The product is called Hemp in Hemp and has only two ingredients on the label: HEMP FLOWER INFUSED IN ORGANIC HEMPSEED OIL. I bottle it in three-ounce maple syrup jars that scream "grown and sold by the farmer." I work on it for 10 months and wholesale it at $50 a bottle.

That wasn't the plan A of my initial group, by the way. We tried to find wholesale outlets for our first crop, in 2016. Then I recognized that with tons of harvested seed and a seed-oil press on the farm, we had a distinct advantage: Almost no one was infusing their flower product in seed oil, let alone their own seed oil, in 2016. That's because most farmers were (and are) cultivating sinsemilla hemp (all-female, from the Spanish for "seedless"). They were seeking flower with 10 percent or higher CBD only. Our hemp variety was a dioecious (male and female) cultivar. These generally contain lower cannabinoid levels than female-only crops, but we harvested those tons of seed.

Today I'm so grateful that I was forced into a farm-to-table product by the wisdom of Dolly Parton and Wendell Berry. So far I've marketed Hemp in Hemp as a muscle, bath, and massage oil. Possibly because of my cultivar's entourage effect, possibly because of the slow infusion mode in hempseed oil, I emerge from a bath infused with ½ teaspoon of it a gelatinous invertebrate.[14] So I can sing the product's praises as its genuine number one customer, which is important. The first small pressing of 750 bottles of Hemp in Hemp, aided by a lot of legwork over many months, eventually sold through.

This is a craft model. Hardly Coors. Vertical integration, as economists describe it when you maintain local management as your raw harvest works its way up the value chain to shelf-ready hemp merchandise, can be difficult to maintain at scale. Though if you do the math, a regionally focused, several-family enterprise that scales up to just 10,000 units at $50 per unit wholesale is making a fine living for its members. And I would argue such an enterprise is much more beneficial to the community where the independent families marketing it live than wholesale agriculture would be.[15]

That's because each step of the production process that an enterprise keeps local both before and after harvesting the hemp magnifies the economic impact on our communities by about three times.[16] This is a very real economic concept called the multiplier effect.

Say a group of farmers cooperatively grows and markets a hemp-flower-based sleep-aid tea at the retail level. The dollars that each co-op member is paid then recirculate in the regional economy, keeping more value on the home front than if the flower were procured from somewhere else. And that's before considering the environmental costs of transportation. The co-op gets its equipment repaired locally (and believe me, farm equipment requires a lot of upkeep), its members eat at local markets, and pretty quickly you see how one dollar spent locally turns into three. "Put simply," said Colin Murray, president of the American Independent Business Alliance, "the multiplier effect creates more local wealth."

Conversely, when someone buys fungible CBD "isolate" (as it sounds, this is CBD that is machine-isolated from the hemp flower) grown who-knows-where for their product, they're obviously helping the hemp economy. But they're not helping the economy in their backyard as much, unless perhaps they hire folks to bottle the product. That anonymous hemp was grown and sold to a commodities broker (a middleman, or in gold rush terms, the shovel salesman) right from harvest. He concentrates it into CBD isolate and sells it to a hemp-product enterprise, for a much higher price than the farmer ever sees.

"Today's farmers get about three cents of every retail dollar from their crops," said Bill Althouse, cofounder of the Fat Pig Society organic hemp cooperative in Fort Collins, Colorado. "Our goal is a hundred cents, less expenses."

The broker couldn't care less about the rural farming economy. To a market trader, price is all that matters. Not soil. Not healthy communities.

Not farmer well-being. Not sourcing regionally. Not humanity's survival. Thus, if there's one overarching message I hope will prove the takeaway from reading this book, it's that the endgame for a thriving enterprise is not buyout by hedge funds or going public. It's regional investment in a farmer-centric enterprise that focuses on regenerative values as a core principle and business MO, from cultivation to delivery.

Easier said than done, one recognizes, but I think absolutely essential if you want your grandkids to have a breathable atmosphere and drinkable water.

We have history on our side. Something I find helpful to keep in mind, especially when some pay-to-play legal team posing as a hemp industry group is proposing big ag-style regulatory standards: We're not reinventing regenerative farming here. We're just having fun rediscovering it after a short break and mapping it onto digital-age society. This is the dawn of the next economic phase that follows "Don't be evil." Maybe we can call it the "Make every single decision in your enterprise as though the survival of humanity is at stake" era.

I love this era. It's already made my own hemp diet—which for years bankrolled the Canadian prairie economy—self-grown and free. In fact, I'm polishing off a hempseed, ginger, and mango shake right now. Sure, it takes years of exhausting work to be a successful regenerative hemp entrepreneur. But trumping everything is that it's just so tasty.

———————

Under the 2014 and 2018 Farm Bills' hemp provisions, each state's agriculture department has to establish a federally compliant state program. Until very recently, this meant my hemp work had to take place in states other than my own. Better than not cultivating hemp at all while we waited out a governor who didn't understand hemp. Now my family is finally bringing it home. The Land of Enchantment launched its hemp program in 2019; and we hold permit number 142.

When the permit arrived in the mail with my name on it, I was a little surprised by how much it meant to me. I mean, I had already been growing hemp for four years. I actually choked up for a second, then shook it off and took a moment to appreciate this genuine triumph. The war on cannabis was done, its legacy a trillion wasted taxpayer dollars and 82,000 citizens still in federal prison for nonviolent offenses. Now I held a hemp-cultivation document in my hands. When I bought the extremely remote, 42-acre ranch in 2005, I hardly dared to dream I'd be cultivating hemp at home in 14 short years. I mean, I dreamed plenty. I believe farming, all (nontoxic) farming, is a human right, but I also didn't fancy an armed raid while homeschooling young children.

Hemp legalization was widely considered a pipe dream back in 2006. Not one member of the US Senate supported it. Today cannabis and hemp legislation generally sails through Congress. And the 2018 Farm Bill hemp provision—introduced by the two most powerful members of the US Senate, Republican Mitch McConnell and Democrat Charles Schumer—made hemp almost, but not quite, as legal as tomatoes.

A long-hoped-for tipping point has been reached. Mainstream candidates now campaign on their pro-cannabis record. I, alongside a few thousand others (and there's room for you), have the immensely fortuitous timing to be participating in the rebirth of a major industry.

On the evening that hemp became legal again, with Julie Andrews temporarily back in the corral (and feeding on hemp-protein meal), I remember switching off the gadgets and strolling by moonlight down to the middle of the near-future hemp field. Besides the usual pleasures of being outside with human and goat kids nearby, I've found I actually have to be in a field to really get a clear sense of how I want to plant, from crop spacing to watering strategy.

Hands on hips, I surveyed the meadow. I would finally get to plant at home, for my family's food. In a few months, I'd be able to see and smell the plants from the ranch house kitchen. Oh, how this made the ol' endorphins flow. It was a primeval feeling. You get a lot of those when you return to farming. The principal cultivar I planned on growing on our small home plot here at 5,700 feet, called Samurai, tested at 31 percent protein. It's a heck of a diet to feed one's family. Talk about "part of a balanced breakfast." That evening, I confess to feeling a touch of exuberance myself.

Part of it is simply that working outdoors makes me happy, or I should say *happier*; I'm generally pretty happy. I have a loving family, good health, and a sweet gig. What's to complain about?

But there's more to it. Our resident great horned owl couple began its evening date (breakfast to them), cooing major third harmonies to each other in stereo above my head. My dogs played tug-of-war with an old cholla stick. In this distraction-free, immensely dense quiet, I felt the particular clarity allowed by a rural life.

The clarity to consciously breathe deeply of clean air. The clarity to know that no far-off government is going to take care of me. And the clarity to know that in any endeavor I'm wise to return an amount at least equal to what I and my family take.

I didn't daydream for long. There was a lot of work to do. The hemp season had already begun, five months before a seed went into the ground. I knew because I had just drawn first blood of the season, compliments of a strand of bear grass I pulled as I weeded a swath of cover crop. Farming tip—blood from any finger is an excellent source of nitrogen for your soil. Sure, the cut was also dripping copiously onto my pants and dogs. Beats a cubicle.

The Farming Year Never Ends Anymore
Vermont and Oregon, 2016

Though the discerning reader will figure this out soon enough, I don't have all the answers about how to best farm and market hemp. In fact, my main intent is to explore whether the entrepreneurial modes I preach can be successfully implemented in the marketplace—particularly when the test enterprise is led by a fellow who possesses neither the door-to-door salesman mind-set nor a particularly green thumb.

My own hemp operation—totaling seven acres across four states in 2019—is just far enough beyond my family's personal use to edge it into the cottage industry category. Not that I'm ignoring the entrepreneurial side. I'm deep into the five-year, slow-growth plan outlined in these pages.

But beyond having a livable climate for my kids and theirs (and theirs), my goal is really just to grow and bottle a product that I enjoy myself. That accomplished, I'd rather be floating down a remote river. The actual catalyst for my, say, hauling tail to Vermont in an ice storm to bottle product or dashing to Oregon to harvest hemp in the wake of a—ho hum—millennial wildfire is I'm tired of pundits on any topic who spout at the mouth but never lay it on the line in the real world.

The effort has left me sufficiently battle scarred to offer some advice throughout the long and intense hemp season. If you are thinking of taking the leap into hemp as a regenerative entrepreneur, here's the first of the five things I wish I'd known before becoming an enabler of the hemp plant's ambitions. For those who simply want to enjoy hemp products, and maybe learn more about what it takes to get them to their store, hopefully seeing what one has to go through will inspire you to hug a farmer.

———————

For a few hundred years, the farmer had a deal with society: When the crop was in, she got paid, and then she was welcome to head inside and "mend harness." (Read: hibernate or, more recently, stream movies and eat popcorn for four months.) But that deal is off in the digital age, at least if you want to be independent and make a long-term living. In other words, if you want to share in the retail value of the crop.

Today your work isn't finished when you're done harvesting your crop—it's just beginning. This piece of advice came to me from my colleague Margaret Flewellen, who founded a company called Natural Good Medicines in 2014. She makes farm-to-table products in Oregon. In fact, I call this truism Margaret's Law, and boy, it sure would have been helpful to know before I dived into hemp.

The 20th-century farmer's work might not have paid well, the middlemen made most of the money, and the "conventional" pesticides were often toxic. But no one was the farmer's boss. And any number of dollars felt a lot more than zero. Even when it was just a little more than zero, after expenses. Someone else turned the wheat into Wheat Thins. All the farmer had to do was get it into the silo. Then it was Miller Time. And none too soon. You never knew whether Mother Nature was going to be friendly in a given year. Backs were sore. Farming was hard work. Today it's the fun part.

When it comes to hemp, Margaret's Law is all the more in play. Not only is a farm enterprise wise to turn its raw harvest into final product, but the enterprise itself is part of the very first expansion team that is creating and defining the markets themselves.

"The farming year never ends anymore," Margaret said on the day she revealed her law to me. "If you're in it for the long haul, you can't just grow it. You have to sell it."

We hemp purveyors, in other words, have the extra job of letting 99.5 percent of the population know that our product even exists. Fewer than 0.5 percent of US households had a hemp product in them in 2017, when hempseed retail giant Manitoba Harvest examined grocery and box store sales, according to Shaun Crew, former president and CEO of Hemp Oil Canada (which has since merged with Manitoba Harvest).

The length of the digital-age farming year is difficult to grasp even for many experienced, multigenerational farming families. Perhaps more so than for new farmers. I know this because I believe my inability to sufficiently convey it was the key factor in the dissolution of my original, 2016 Vermont partnership.

John Williamson, 57, was one of the best farmers I'd ever met. I probably learned more about the metrics of large-acreage farming from my collaboration with him than I have from any other human before or since. Along with a third family of good folks and fine farmers (Robin Alberti, Ken Manfredi, and clan), we planted 23 acres together on his third-generation farm outside Bennington.

John understood soil and processed biofuel in his barn so our combine harvest was petroleum-free. At planting time he even welded a roller extension onto our sowing rig so that seed-soil contact would be sufficient for germination every-place we dropped seeds. I've carried innumerable lessons like that one into my ensuing hemp projects, and taught them to others, from new partners to academics to consulting clients. The guy was just a quality human being. Our 2016 harvest remains a benchmark for beauty and productivity. We brought in 1,000 pounds of seed per acre on our most productive fields, with that 31 percent protein in the seed meal and a lovely, terpene-rich flower harvest to boot.

John had been willing to test the hemp waters because the low prices his alfalfa crops were demanding made it "barely worth planting anymore." But, as crappy as the associated dairy feed market was (and is) in Vermont, at least he used to get *something* come October. Enough for popcorn. Maybe enough to build a barn extension. So when I told him that not only would we not immediately be paid the moment the combine had deposited our bounty of seed and flower in various silos and storage bins, but that additional time, funds, and equipment were needed for months, perhaps years, he was out before the following spring. Alfalfa prices might not be tenable, but zip is a scary number to absorb for someone who has just burned rubber for eight months. The idea of creating a product that we would have to bottle, label, store, and peddle was anathema to him.

When things first came to a head, I was puzzled. I thought I had explained to the initial Vermont team that multiyear endurance was a prerequisite; that markets were so immature that we basically had to describe—at trade shows, farmers markets, and food co-ops—what we were even offering. Evidently I hadn't explained clearly enough. In retrospect, I suppose I had dangled the possibility that wholesale prices could be high enough to justify selling our seed and flower right at harvest, the way John had always sold his alfalfa and other crops. Come harvest, I realized our best play was to combine the seed and flower into a value-added massage and bath oil.

Part of me wishes there had been a functioning wholesale market for that first harvest in Vermont. But even though it cost us that partnership, on balance I'm glad that I had to learn the lesson, because I am very proud of the product that resulted. I was forced to listen to Dolly Parton. And I think that a fine farm-to-table offering has resulted. At the time, though, there were some tense moments when I realized that the "well, now our work is really beginning" message wasn't getting through to a lifelong farmer.

A year or three is a long time to ask someone not to deposit a check. This is when things get real. This is when partners forget previous conversations, because, say, an aunt needs a medical procedure or a transmission has just dropped off in the middle of a cow patty. All I can suggest is that you try to be as prepared as possible for the reality of early hemp multitasking. Draw up a multiyear budget and try to stick to it. And recognize that you can't just harvest a crop, you can't just create a cool product: You have to create a market for it.

I keep seeing this early Vermont lesson repeated in my colleagues' enterprises. It comes down to, "choose your partners carefully and lay out expectations before you even create your entity." Janel Ralph, founder of Palmetto Harmony, a CBD company in Conway, South Carolina, told me she turned down half a dozen partnership offers over the course of five years before saying yes to perhaps the world's most accomplished hemp entrepreneur, John Roulac, founder of the California-based company Nutiva.

After two decades spent engaged in the pleasantly solitary act of writing for a living, I find that the hardest part of the whole independent-farming renaissance is dealing with other people. And I'm sure those other people can say the same of dealing with me. It's easy for me to see now that I share equal responsibility for the dissolution of that initial group. At the time, of course, knee-deep in empty bottles and disgruntled colleagues, I felt as if I were being abandoned. That was when,

3,000 miles away, in Oregon's Emerald Triangle, my West Coast pardner Margaret Flewellen laid the hard truth on me.

"Oh, you didn't hammer home that farming is year-round now?" she asked. "It's like teaching a new language. The more experienced a farmer is, the harder it is to convey. The old mode is in their blood. You can't just tell them once, or three times. That was probably your mistake."

I won't forget Margaret's Law again. Margaret, 48, should know. After years as a medicinal cannabis provider, today she crafts her hemp products from her family's Oregon harvests: She's got ethanol processers for tincture; fancy, giant-screened computers for labels and web marketing; and she even owns her own cellophane-rolling machine. She uses that to seal the hemp cartons for her line of Zenith CBD "hemparettes," a product aimed at helping folks kick tobacco. That zaftig machine lives in the guest room where I sleep when I'm in for fieldwork—Margaret and her husband, Edgar Winters (this is not the rocker Edgar Winter of "Frankenstein" fame, though they are third cousins), and I have teamed up on genetic development and consulting.

She passed her law on to me, so now I hope the message helps you: You are now and forevermore an entrepreneur. And hemp entrepreneurship is almost definitely going to require multiyear endurance before you see light at the end of the workload and revenue tunnels. As 68-year-old Edgar puts it, "I work each day until I'm asleep on my feet. I lie down for a few hours. And then I get up and do it again. Doesn't matter if it's June or January. And I really don't see any other way."

Then he adds, "And when you're really ready to drop, this is when you get a call from someone wanting an hour of your time to ask you how it's done."

A week off every now and then would be nice. But "We'll sleep when we're dead" is something of a mantra for Edgar,

Margaret, my sweetheart, and me. And yet most of the time you'll find me smiling and up for a swim. Strange.

Some professional farmers who are used to selling a crop of any kind to wholesalers will make the entrepreneurial transition, and some won't. They definitely are a high-risk group for quitting after one season, like silent film actors who weren't able to acquire the skill set necessary to survive in talkies. On the other hand, some of my hemp colleagues simply love the never-ending challenges.

"I'm having the time of my life," William "Wild Bill" Billings of Colorado Hemp Project told me the other day, and I think he was at least 75 percent serious. The 69-year-old's cell phone was in one ear as usual as he closed another seed deal somewhere far away.

Wild Bill, as everyone calls him, from US congresspeople to Jamaican ministers, was part of Colorado's first federally permitted hemp crop in 2014. Fiber from that harvest went into an insert included with the 25 percent–hemp monograph I wrote about the 2014 hemp season, called *First Legal Harvest*. Now Wild Bill has got his fingers in hundreds of acres all over the world. This is a guy who knows where his work gloves are. Hemp is his calling. It has to be.

"It's a new world," he told me. "Farming is gonna save us. But it's not your grandmother's farming."

There is one connecting thread between today's farming and your grandmother's farming, though, and it's a fundamental one: the soil. As a producer, once the question evolves from only "Can I make a living?" to "Will my hemp enterprise also be of value to my community and the atmosphere?" the answer begins exactly there: not in the lush green plants to come, but in the decisions you make underground long before the planting season even starts.

CHAPTER TWO

We're All Soil Farmers Now

Colville Tribal Land,
near Omak, Washington, 2017–2018

*There's always as much belowground as above . . . mostly
unknown microbes and invertebrates, perhaps a million
species . . . [including] fungi that infuse into the roots of
trees in partnership so tight it's hard to say where one
organism leaves off and the other begins.*

—RICHARD POWERS, *The Overstory*

Wwhat goes on under the surface of your farm before
you plant determines what springs forth from that
ground. Simple as that. Your underground work is
particularly vital if your soil, like much of the planet's farmland, isn't
currently in tip-top shape. The first six inches below your feet are
especially important, according to most soil experts. But this depth
will vary by ecosystem.

Hemp is grown on every continent except Antarctica, and in every
soil type, but no matter where you cultivate, if folks wind up opening,
pouring, and enjoying a bottle of your superior hemp product, it's
going to be because you took the time to work on your source soil. Soil
is Step One.

Because every plant that grows outdoors lives in a soil home, making
that home a pleasant and healthy one is a no-brainer. What is astounding

is how quickly awareness of soil health primacy dropped, in one century, from 100 percent to maybe 1 percent among worldwide farmers, if you measure the percentage of land cultivated organically since the "better living through chemistry" era began in the 1930s.

Even as repurposed nerve gas was starting to be used as insecticide and petroleum turned into nitrogen fertilizer, plenty of top-level of US deciders were aware of the importance of healthy soil. The Dust Bowl brought Senate hearings on the subject, and President Roosevelt wrote state governors in 1937, saying, "The nation that destroys its soil, destroys itself."[1] At the policy level, then chief of the USDA's Bureau of Chemistry and Soils, Charles E. Kellogg, wrote in the department's *Yearbook of Agriculture 1938*, "There can be no life without soil and no soil without life; they have evolved together."[2]

Building soil is a step much of humanity's farming community has chosen to forget for a little while. Today most farmers have adapted to a cultivation sequence that includes what Edgar Winters calls "nuking the soil" before planting a crop that is designed (more than bred) to survive in a nutritive wasteland. If you don't care about taste and maximum nutrition, that works for a little while, until herbicide toxicity or super-weeds creep in. Nat Bradford, a fifth-generation farmer in the American South, told me recently that "Under today's rules in our state, it's dang near impossible to certify a seed of any kind without using chemical herbicides." Yikes! Carcinogens are not a control group.

Seed quality, weather, water availability, markets—none of these matter until you take care of the largely invisible underground ecosystem. We might call ourselves wheat farmers, hemp farmers, or backyard gardeners; in actuality we are all soil farmers now.

With crop-yield decreases across the food crop spectrum and across the planet, there are only folks who realize they are soil farmers and those who soon will. If you realize it, there is hope for your crop and our species. But you might be wondering where to start on, say, your cousin's Missouri back 20 that you're trying to save from becoming 10 units in another McSubdivision. Let's assume the field has been recently planted with GMO (genetically modified organism) soy and associated inputs like Monsanto's glyphosate, for a short while longer the world's most widely used herbicide.[3] What's the first soil-building step?

Mushroom Hunting in Bigfoot Country

Mushroom hunting. That was my job one mid-April day in 2018. A half dozen colleagues and I were foraging for white veins of mycelia on Colville tribal land east of the Cascades under the guidance of a soil-building expert named Chris Trump. We were investigating the possibility of supplementing the tribe's already intensive second-season soil-building effort by employing a technique called *Korean natural farming* (KNF, sometimes just shorthanded as *natural farming*). My hands were black with humus before lunch.

The reason we were muddying ourselves was because Jackie Richter, the Colville Tribe's hemp project coordinator, knew that soil was a mission critical step in the hemp year. The tribe had helped launch the modern era of Native American hemp the previous season with 60 acres, and scaled up to 105 for 2018.

Jackie had a track record of agricultural success with the tribe. She'd built seven-figure-earning fruit orchard projects, even while running her family trucking company and, I noticed, making it to her son's baseball games. This is not a quitter, and she learned long before we met what it takes to make a commercial crop succeed. I had a very close view of her work ethic: I was lead consultant for the project for nearly three years. And when she was in the field, you would often find Jackie stooped over, digging into the soil, or gathering samples for nutrient testing.

"We've always had a holistic management plan for this project," she told me when I first introduced her to soil expert Chris Trump. "After years of conventional farming, I know there's a lot of regrowth that needs to happen in this soil."

Everybody on the core team of this first federally permitted Native American project was aware that soil health was the key area where the effort needed improvement. When the project launched on the banks of the Columbia River in 2017, the Colville field wasn't what anyone would call optimally robust soil. The finest stand of hemp from the 2017 debut crop, by far, had been a narrow, several-acre panel where the project's principal farmer had run his cows the previous year. In other words, the crop had grown best in the one place where there was a healthy dose of nitrogen. And that was just one nutrient in need of replenishment.

So Jackie, who was working closely with a regenerative-soil-building expert named Erika Winters, thought, what the heck, let's also bring in this KNF fellow and hear what he has to say. When Chris Trump arrived on the scene, planting time was less than six weeks away.

The tribe had decided to apply as one of the very few entities to wade into Washington's immensely flawed initial hemp program. Under the often sleep-deprived leadership of Jackie, the project was bravely forging ahead and nearly doubling its 2017 acreage.

That meant soil-building work. Mucking up my hemp version of Carhartts (a brand called i.N.i Cooperative, grown in China), I asked Chris Trump why we were starting with fungus-tracking. "Much of our depleted farmland is totally devoid of fungal life," he explained as we climbed from the Colville field into a dreamscape pine-and-fir hillside above the river. "And hemp, like many crops, likes a balanced fungal-to-bacterial ratio."

Fungus, in other words, is just one component of the immensely diverse, if tiny, nutrient forest you're wise to nurture even before you plant your main crop. And, sure, it's a critical one. But your soil is an ecosystem as complex as any tropical rain forest. Your subterranean "neighborhood" ideally includes microscopic residents such as beneficial bacteria, protozoa, yeasts, and nematodes, to name a few. And don't forget earthworms and a whole slew of visible denizens of the soil's root zone. All work together with a shared goal: making nutrients available to the roots of your hemp and thus your eventual hemp product.

Beneficial fungi are where you'll want to focus your first efforts. They are the key homeowner group in your soil neighborhood. Collectively, we'll call them *mycelia* during the vegetative phase when we gather them. Many of us who delve into the fungal life cycle have that moment when we really recognize that, just as we were taught, fungi aren't little plants; they are their own kingdom. Indeed, mushrooms are much older than tall forest trees. But broadly speaking, mycelia clusters are analogous to plant roots, and the visibly sprouting mushroom is akin to the flower.[4]

It's no exaggeration to say that mycelia are a key part of your hemp bottom line. Your goal in the field is beautiful plants. Every hemp farmer surges with pride when her farm is bursting with pungent flowers, each testing gravity with its crystalline trichome frosting. Trichomes are the tiny, sticky cones on female cannabis leaves and flowers where the canna-binoids like THC and CBD reside. When you get very close to a hemp flower (and I recommend you do, often—just watch out for bees), the trichomes look kind of like translucent amber mushrooms themselves. Even when you're growing for seed, you still love to see dense flower formation: It's visual and fragrant proof that you're getting the job done in the field.

That's the goal, come fall. But on that day in April, the Colville tribal field was bare and in need of help. This is why we were training our gazes to focus on very small critters. We weren't seeking morels or chanterelles as we tromped in ankle-deep mush, but little wavy white trails of fila-ments that are found thriving under leaves and rocks and alongside roots in the liveliest pockets of your watershed's forest ecosystem.

"The best places to look are directly along the source of water flow to your fields," Chris Trump explained to me as my boots suctioned flatu-lently out of a bog. "By gathering the mycelia from the forest above your fields, you're transporting them to a soil that they love and recognize, and in which they can thrive."

This remains one of my favorite parts of conscious soil building: that, for best results, you want to gather your nutrients close to home, a concept agronomists call a *closed nutrient loop*. And I love even more that the fungi you unearth for your own field are likely not found anywhere else on the planet.

The plan was to find and gather these blooms and mix them with rice and sugar, thus expanding the colony into a rich compost tea before diluting it and applying it to the tribe's field. Trump, 36, is a fellow known in soil circles for bringing this technique to large-acreage farming, hav-ing proved that it works on his family's 800-acre macadamia nut farm on Hawaii's Big Island. He's since moved to Idaho, which is another reason why he was a perfect choice for schooling the comparatively nearby Colville project.

Chris Trump's professional life is dedicated to soil regrowth. I don't think I'll ever digest everything he explained to us in those hills. I was constantly jotting down the more polysyllabic bits of microbiology dia-lect to look up later. As the Alabama-based soil expert Michael "the Dirt Nerd" LaBelle once said to me, "The interplay of microorganisms is not rocket science. It's much more complex than rocket science." But Chris's key theme was that fungus is something of the glue, the undercurrent, the foundation of the microbial pueblo.

I was his willing student, and not just because I was on the clock for the Colville project or because the information might help my own entrepreneurial hemp projects that year in Oregon and Vermont. I was thinking about human survival. Specifically, I was thinking about those three billion tons of carbon that each inch of restored topsoil sequesters annually. Three billion tons mitigates 10 percent of annual fossil fuel carbon emissions.[5] That's just one inch of soil. Imagine four or six, worldwide. That much provides us a window to finally realize, en masse, that the sun's power is free, and to rev the biomaterials-based economy into gear.

The main reason hemp is such a soil-building aid is largely mechanical: It's not that hemp itself is a nitrogen fixer like alfalfa, clover, or vetch, but rather that its strong taproots create a transportation network for your soil's beneficial microbes. When thriving, these microbes allow your plant to perform its photosynthetic gymnastics like an in-shape athlete.

Soil microclimate building feels particularly timely in Eastern Washington. According to the Washington Grain Commission, the state's wheat country produces more of the amber grain than Greece—2.2 million acres.[6] I started the day driving through a swath of it. Wheat, wheat—everywhere I looked for hours was just a landscape of monoculture wheat, most of it sprayed year after year with herbicides such as dicamba and 2,4-D. You could fly over this land in a small plane for an hour and not see a break in wheat acreage.

For a century this region's soil has been a key breadbasket of the nation (or at least a Wheat Thins basket). You can't ignore wheat if you live north of Spokane. Its seasons dominate everything. But Washington's wheat represents monoculture at the tail end of its peak, with yields stagnating or declining as soil weakens. Whenever I've paid Colville field visits in between wheat seasons, the landscape looks like the Sahara.

Change is on the horizon. Lifelong rancher Richter told me, "Wheat farmers tend to be cautious folks who have done it the same way for a long time. Now they're seeing wheat hit an all-time low in prices. And they are definitely paying attention to what we're doing with hemp. It'd make a natural transition. They want to grow what makes their families a living."

Korean Natural Farming

"Look for the filaments we can see in strands," Chris Trump directed us as we trudged through a streamside cluster of ponderosa pines, mullein, and willow. The season's first wildflowers were just coming into bloom, especially the creamy white arrowroot, and at one point during the trek we had to sidestep wild horses.

Chris is a solid fellow in spirit and build, who speaks in the kind of near whisper that makes you pay attention, especially outdoors. And there in the Columbia River basin, he was usually describing important things, such as the difference between beneficial and potentially dangerous fungi. He referred to the colonies like old friends. "Under fallen wood and leaves is best, because that's what these guys like to chew."

Our group of microbe trackers included Trump, two of his interns, Jackie, myself, and the married couple who wound up buying the entire 15,000-pound Colville seed harvest seven months later, Gregg Gnecco and Tonia Farman of a seed-processing start-up called Hemp Northwest. We struck white gold after about 35 minutes of hiking, and tucked a pound or so of our crusty booty into glass jars. More than enough to make tea for 100 acres. I asked Trump how it is we can actually see the strands, since what we're looking for, as individuals, are microscopic.

"When they're collecting in clumps larger than four micrometers, they become visible to us, and at that size you know they won't be dangerous," he said. "We're aiming to get a snapshot of the local beneficial species and take 'em down the hill to the farm."

On our way back to the acreage late that morning, I spun around in camera-ready position with every sudden sound, whether a startled hawk's flutter or a colleague's water bottle clanging on a carabiner. That's because this area of moss-covered second growth and dense huckleberry clusters 100 miles from Canada is the prime spot for snapping that $100,000 Bigfoot photo for which the *National Enquirer* surely has a standing order.

I have several questions for Sasquatch, including, "Is there a whole Bigfoot-ed community, does it desire a hemp consultant, and is there internet access?" But before I could sneak off to make any kind of coordinated stakeout, Chris called me over to show me a particularly impressive tangle of microbes.

I first learned about KNF at a Hawaii Farmers Union convention workshop in 2016. Folks on the Big Island pretty easily convinced me that it was legit. KNF comprises a fairly complex and formal regimen of soil development. There is a catalog of soil-building steps when you really get into it, all of them identified with abbreviations, such as the IMO (indigenous microorganisms) step, the WCA (water-soluble calcium) step, and the FPJ (fermented plant juice) step. FPJ, Trump said, is an enzyme-rich concoction aimed at providing plant energy.

All soil theories worth their salt—including KNF—are based on the premise that what you think of as your final crop starts with a mini version in the soil. The soil-building regimens they propose guide you to making a home for the microbes. In fact, Chris suggested that the tribal project conduct a special soil test of not just traditional N (nitrogen), P (phosphorus), and K (potassium) but also active bacteria, fungi, and protozoa. Which about sums it up—the teamwork amongst these families of critters gets you going with microscopic soil building. I noticed when I checked the soil-testing company's catalog that testing for nematodes is extra. That's a sentence I never imagined I'd write.

Just as Chinese medicine and Western medicine take different approaches to the shared goal of good human health, so Korean natural farming is about a microbe-focused process that, the theory goes, leads to more bioavailable nutrients for your crop than dumping synthetic nutrients out of a bag.

Bioavailability describes the proportion of a mineral, vitamin, or sugar that a person or a plant can make use of from its diet. It's one of my favorite nutritive concepts. We're talking about the reason building soil provides a competitive advantage. Forget apples and oranges. Even talking about apples and apples hinges on whether the nutrients reach your body when you take a bite. When we discuss bioavailability in the soil, we mean maximizing the nutrients our plants can absorb in their roots and in turn provide us with in a supremely dank hemp crop.

To get your best hemp crop, you're aiming to create the conditions that allow all these kingdoms of microflora and -fauna to form the kind of neighborhood we all wish we had in our macroscopic lives: Everyone thriving in harmony with everyone else, providing a cup of sugar when asked, making extra room if desired.

"Just as in a human community there might be a blacksmith, farmers, and a nurse," Trump said, "we are nurturing a microbial community, which occurs by itself in nature, but which needs rebuilding in most of the world's farmland."

The result for the farmer is that a healthy amount of the traditional nutrients we Westerners think of as valuable for our plants (good ol' N-P-K) is getting to the plants.

"Ultimately, a plant's roots trade sugar for nutrients," Michael LaBelle told me. "And microbes in your soil act like your gut preparing your stomach for food."

A takeaway here, I hope, is that you're wise to test your soil before you start growing hemp. But soil nurturing doesn't necessarily mean adding a huge amount of amendments to your crop. Even as you learn as much as you can about your soil's starting condition, your soil building can include a strategy of not doing much. Some of my most thriving hemp crops to date have sprung from soil that had been fallow for 10 years or more, and on which we human farmers did little soil amending.

I find I'm very good at not doing stuff. Ask me to skip a step in any process, and you'll find me profoundly adept. Heck, I've left goat poop and alfalfa hay tailings sit so long in my truck bed that it sprouted fresh alfalfa. Still, not wanting to rely on my mere real-life experiences in the field, I checked with the Dirt Nerd LaBelle to see if he agreed that sometimes less is more.

"If your soil is pH-balanced and you lay the basic groundwork, you can let the microbes do the work," he said.

"Well, I've got twelve years of goat Milk Duds on this soil."

"Then you're golden. If nitrogen is decent on your farm and you can add only a few things—in most cases, it will be calcium and magnesium— you will be on your way to a healthy crop. It's usually trace minerals that are lacking. That's what you should think about supplementing."

And that is exactly our strategy on the Funky Butte Ranch this year. We're starting with mycelia, some overwintering nitrogen-fixing cover crops, and goat poop–alfalfa mix. Plus some diluted kelp. That's it.

Modern KNF was founded by Cho Han Kyu, 84, today called Master Cho by his fans, who blended traditional Korean cultivation styles with modes he learned in Japan. It is one among an array of clever techniques that humans across the world have discovered over millennia to ensure that the same, limited amount of soil is fresh and ready to go again next season. No one technique represents the only right way. Although I'd be remiss if I didn't mention that I visited a comparative trial on Maui, where a plot of KNF kalo just crushed the "conventional" control crop.[7]

I tend to lean toward applications of any kind, in any field, that are based on processes that have been functioning for 8,000 years. I like long-established farming techniques because most of human history has taken place before what we might call the Whole Foods Era. The farmer's task until very recently has been simple and stark: Grow food successfully through harvest and storage, every single year, or die. And humans are still here.

Take Korea: Korea had pretty much only Korea's fields to rely on for food until the Whole Foods Era. Folks there, rich or poor, didn't have Chilean *robo*-salmon to import or Mexican takeout around the corner. So perhaps eight millennia ago, they figured out a regenerative way to manage the same fixed acreage every spring.

Mapping Trump's lessons to my home crop, the reason we're actively building soil—including unintentional if regular supplies of my own blood—is because it had been overgrazed by mules in the 1990s, long before I moved here in 2005. For the most part, though, my theory is, "Create a desirable home and let the soil do its thing."

The Costs (and Benefits) of Soil Building

When I asked him about the various costs that our putative 20-acre hemp farmer would need to consider to build soil according to the natural farming system, Trump said that to start you can expect your input and labor costs to be about $100 per acre for the whole KNF regimen. That's equivalent to the start-up costs of "conventional" farming.

"But in Hawaii," he said. "We've learned so much over a decade and we've got the system locked in to the point that our macadamia nut farm costs are down to twenty-seven dollars per acre." By contrast, Jackie said the Colville project's soil-building costs for the Colville crop in 2018, which primarily used standard organic inputs, was $23,000. That's several times higher than Trump's per-acre cost. One of Chris's key educational bullet points is that as your soil strengthens, annual costs go down. In fact, he pointed out that one of the most appealing elements of natural farming is that by encouraging local microorganisms, they can begin to reproduce on their own, reducing input time in each successive year.

"We do about three applications [of the various KNF regimens] per season now, when we used to have to do ten," he said.

Chris Trump works full-time on his soil. For those with less patience, plenty of folks will gladly offer you miracle soil-cure formulas (what Chris calls "bugs in a jug"). These are the gold rush shovel merchants. If you do your research, I'm sure many of these products are terrific. Plus, I'm glad farmers are talking about soil microbes again. But Chris believes that lasting soil building, like success in the wider hemp industry, is not a quick-fix proposition. What it is, is potentially enduring.

We returned from hillside mushroom picking in the early afternoon. As we set up shop for the cooking stage of fungus building alongside the plowed-and-ready Colville field, one of Chris's interns, Morgan, set a centerpiece of huge, delicately white-petaled arrowroot fronds on a project truck bed. Beside it, we mixed up the concentrated base of a mushroom tea to feed soil that would then feed a hemp crop.

"Fungi are the most finicky microbe to cultivate," Chris said as he pulled out a rice cooker—a device you don't see in a hemp field every day. "In nature it takes an animal to transport them and drop them off their hoof, or something like a landslide. The bacteria and yeasts are far easier to grow out."

So fungus breeding it was. The first thing we did that afternoon was cook up three pounds of organic rice, the medium on which, after mixing in some sugar, we further bloomed our fungus colony.

"Rice is among the fattier of grains," Chris explained. "The fungal colony chews on the fat-carb mixture. For now, that's its nursery."

With the mycelia feeding, Chris showed us how to transform the resulting concoction into a shelf-stable, dormant state. This involved drying the mixture thanks to the chemical bond created between the mixture's sugar and any water. "This part is like salting fish so it won't rot," he said. "While it's stored, it's essentially in a cryo-freeze."

When you're ready to spread it in your farmland, he said, "Grab the jar from your shelf, dilute it into tea form, and you have this indigenous, ready-made inoculum that will benefit your crop."

Chris went on to explain that there's quite a bit more to be done than he could cover in one afternoon before a compost tea is ready for soil application. He outlined four steps, in fact, which are called indigenous microorganisms 1 (IMO1) for the first step, which is that rice mixture we made, IMO2 for the second step, which is the shelf-stable phase he'd begun to demonstrate, and so on. Experienced KNF practitioners are meticulously detailed about each step in this process, which at later stages involves mixing your fungal culture with soil to commingle microbes, and bringing in the nematodes and microarthropods—the next level of the food chain. Then (condensing several steps here for brevity), you "brew it out" to get your concoction ready for application to your waiting field. You can do this on small or large acreage, by making and applying a batch of compost tea from a final-stage IMO rice culture at a rough ratio of 20 gallons per acre. For large acreage you add the tea to watering lines. For smaller gardens, you hand-apply it to soil.

As much as I liked what I was seeing that day in the Colville field, equally strong confirmation came in the form of a lifelong professional farmer's buy-in. The Colville project's man on the ground was 58-year-old Dan Townsend, who joined us for much of the mushroom-focused afternoon. He was the professional farmer, the guy who knew how to repair a combine and fix the massive tube network of an irrigation system known as a *pivot*. And he came to hemp recently, from a family with three generations of wheat experience.

"Creating this kind of living system in the soil, I can see straight off that it's likely to work," he told me that spring evening as we closed the farm gate. "It makes sense to me, and I definitely would want to play with it some more."

The Most-Improved-Crop Award

Access to the Colville hemp field comes either via a tribe-owned, state-operated ferry or through a spine-readjusting network of Forest Service dirt roads for which FOUR-WHEEL DRIVE ONLY isn't a strong enough warning. For three years, I had to explain to the Spokane rent-a-car folks each time I returned a crumbling, not-ready-for-prime-time "crossover" SUV that I had been out of the United States for most of my visit.

"You mean Canada?" the counter woman usually asked, contemplating what kind of surcharge I might be subject to for moving into kilometer territory.

"No, tribal land," I replied. And that usually ended the discussion.

The remote field access meant it wasn't so easy to bring even regional experts like Chris Trump on-site to help the project. The reason Jackie, who operated on a chillingly tight budget, was willing to take us fungus gathering is that she understood that soil is currency. "Everything, every hemp shoot, every fruit on every tree on every farm, starts with soil," she told me.

That's why this modern farmer once again, like her farmer ancestors, isn't looking first and foremost at what's happening above the ground. She knows the future of any crop can be foretold by an ecosystem that is otherwise too small to see.

For his part, Chris Trump firmly believes that the difference in quality between the harvests from 20th-century-style conventional agriculture and conscious soil farming will become so large that simple crop performance will propel what is today considered alternative into tomorrow's mainstream.

"As more farmers regain soil-building knowledge," Trump said. "I've come to believe that in the near future we'll see a complete cultural change. The food on the shelves will be profitable to the farmer and, by being healthier, will reduce the runaway medical costs in society."

By the time of Chris Trump's visit, it was a little late to say what role a bit of KNF work would play in the 2018 crop. As a result, the KNF lesson was largely educational, something to keep in mind for future seasons (and in my case, for other hemp projects around the world). We all learned a lot that day, but planting was upon the project, and there

Soil Building, Slow and Fast

Our presumptive farmer trying to save his cousin's Missouri back 20 might want to know how long it will take to transform soil from stressed to healthy. That's a tricky question, because even fields that are a few miles apart can be composed of distinct soil qualities. Some hemp farmers tell me that their baseline N-P-K samples have improved markedly after one season of hemp combined with organic inputs, and a few tell me that at the end of their third season of planting on stressed soil, they haven't yet seen much improvement.

One of my soil-building teachers is Vincent Mina, who operates the six-figure-grossing Kahanu 'Aina Greens, an organic microgreens enterprise, from his small home farm on Maui. (If you're on the island, try the sunflower and radish greens especially.) The 65-year-old is always inviting me to stick my arm in his man-high soil pile, and insisting he's a soil farmer first, not a food farmer.

"This chocolate cake," he told me recently, pointing to the pile, which was full of crawling earthworms, "is everything. This is our business. The microgreens come from this pile of soil."

In recent years, Vincent and I have been talking about collaborating on a hemp project, and he suggests we spend at least three seasons planting soil-building cover crops before we even consider a commercial crop. He should know: The Mina family began building its soil in the 1990s, and Vincent said it took more than five years to reach a place where he felt comfortable with the resulting product. But I've also witnessed soil transformation unfold surprising quickly, on the same hunk of Maui volcanic rock.

Forty-eight-year-old James "Kimo" Simpliciano was a longtime resort chef in Hawaii, until he noticed that it was nearly impossible to serve traditional canoe crops, like poi and breadfruit, at the commercial level.

"Everything was shipped from the mainland or Asia," he said. "You could hardly find local salad greens."

In 2013 the Filipino Hawaiian local decided to become a farmer. He formed Simpli-Fresh Farms, got a grant, and settled on a spot above Lahaina that had been such prime farmland during Hawaii's pre-Western period that the king could look up from his front lanai and check on the well-being of Maui. The problem for Kimo was, for the past 150 years, the spot had been used to cultivate one crop: sugarcane. The soil was so compacted that it came apart in chunks you could bowl with. Its color was a wan beige.

Fast-forward two years. A friend took a picture of Kimo and me doing yoga high up on a parched, unplanted hillside of sugarcane detritus above Simpli-Fresh Farms in June of 2015. Where we stood frozen in Tree Pose was a human-created desert. But in the background of the photo, hundreds of yards below us, the demarcation line couldn't be missed: Where Kimo's expanding polyculture remediation work had brought this land back to vibrant life, the soil was dark and full of plants, alive. Papaya windbreaks, waving bushes of moringa, cacao—they were right in front of our eyes, and undeniable. Talk about a line in the sand.

Healing happens. And faster than Kimo could have hoped. Unfortunately, the story has a sad ending, though only temporarily. In 2018, a—ho hum—devastating millennial fire burned the whole thing and Simpli-Fresh Farms had to move. But the evidence remains. I keep a photo of this nutritive morality tale on my laptop desktop. It tells a story of two possible futures for humanity.

was not enough time to implement the full KNF game plan. We sure got fired up about fungus, though.

Using some of Chris's ideas but mostly those of Winters, Jackie, and Dan, the second-year Colville crop was one of the most beautiful I had yet seen, certainly in the large-farm category. And without a doubt, it won the most-improved prize.

Of course, harvesting a lovely crop and marketing it are two different processes. But Jackie's stated goals for season two in Colville were: (1) Grow a gorgeous crop and (2) sell it. And she did both.

At the end of that KNF crash course, as I did every evening on the Colville project, I stripped down and leapt into the Columbia River. Then I got dressed, descended the dirt track from the hills, and took the ferry back across the same river. Then I climbed out of the Columbia basin, wrote questionable poetry about the sunset, and drove through another 2.4 million acres of wheat on my way back into the United States.

I'm a full-blown fungus advocate now. In February of 2019, with the days getting perceptibly longer on the ranch, my sons and I chugged up the 6,000-foot-high Land of Enchantment arroyos as we always do when lollygagging on the day's indoor work. This time, though, we brought our sampling bags (repurposed paper sandwich sacks).

We gathered what we hoped were sufficient local mycelium strains from under the roots and rocks of our wilderness backyard ecosystem. We stored the rice-sugar-'shroom combo for three months, until the frost danger passed. Then we mixed it into our field alongside a few hundred pounds of alfalfa-encrusted goat poop.

With the ranch's lavender and currants budding and the rhubarb pretty much ready to pluck, my mouth was watering that day, just thinking about the coming season. My heart was pounding, too, but that's because I was a little shaken from the moment, high in the Funky Butte boulders, when my oldest leapt out at me, shouting, "I'm Bigfoot!" I got the photo, though.

By building soil for months before your product is on shelves, you're manifesting the Prophet Dolly's dictum in the "better" category. If we can collectively get the word out about the nutritive bioavailability and climate change mitigation in our mode of cultivation, we independent farmer-entrepreneurs are going to have enough of a customer base to do better than just fine. All by choosing the most regenerative, patient mode for soil building.

Your Industry Starts on the Farm
Fort Collins, Colorado, October 2015

Be on the farm. Every day, ideally, especially if you're one of the primary farmers. But be there at least every week. No matter your title in the enterprise, consider yourself a farmer. Put "hemp farmer" as your tax return occupation. There is no role in the operation that should keep you off the farm for long. Not logo designer, delivery driver, or traveling trade show booth roadie. And certainly not president of the company. You work for a plant. Keep a close and loving eye on it.

In October of 2015, I made an early field visit to the Fat Pig Society hemp worker cooperative (FPS) outside of Fort Collins, Colorado. It was a chilly evening in the Rockies when I arrived close to midnight in my camper and parked beside the co-op's 10 acres of flowering hemp. In the early moments of the modern hemp industry, I had come to learn from the only organic farmers I knew.

FPS cofounder Iginia Boccalandro met me at the rig to see if I needed any extra blankets or snacks. Before wishing me good night, she noticed my alto sax case on the passenger seat and assured me that impassioned jamming on the porch of the co-op's farmhouse at 6:30 a.m. on a weekday in the middle of harvest season would be well received. I'm still not sure if she was kidding. Either way, she helped solidify this "be on the farm" lesson for me.

Iginia is the buoyant, 54-year-old right brain of the FPS organization. It sums up her personality to point out that she's a former Olympic luge racer for—and this is not a typo—Venezuela. She says everything with saucer-eyed exuberance, as though we're all about to take an 85-mile-an-hour sled run. She's also one of my principal hemp mentors to this day.

"Bill's in the kitchen by five thirty pounding his third pot of coffee and scanning the headlines on his laptop," Boccalandro said, referring to 63-year-old Bill Althouse, the brilliant and cantankerous engineer, geneticist, and left-brain energy source of the FPS.[8] The rest of the co-op's four original members were asleep when I pulled in, so my morning entrance would be the first Bill would see of me this visit.

It was Bill who had named the co-op, following the six years he spent working as an electrical engineer in the Pacific island nation of Palau. There, a fat pig in the village means everyone's doing okay; if the pigs are plump, it means the people are, too. And it might be time for a potluck.

I was fired up. When I woke, I would be helping to harvest my first federally legal outdoor sinsemilla (CBD) crop. Even better, this harvest was taking place at what is still one of the few (hopefully not for long) organic hemp cooperatives on the planet. I'm a fan of cooperatives, especially for their focus on local economies.

In 2015, the world was reawakening to hemp. A token 220 acres had been grown nationwide the previous season under the new 2014 Farm Bill provision. By the time I rolled into Fort Collins, it had already increased 1,000 percent to 2,200 acres. Althouse had been waiting his whole adult life for this moment. Though after a half century as a film grip, Western swing dancer, and fruit and decorative flower farmer (these filled the gaps in his engineering résumé), it might have occurred to me that he could use his beauty sleep.

"Bill loves jazz," Iginia said before crushing me in a good night embrace. "Do you know any Coltrane?"

"Um, I can play 'My Favorite Things,' kind of," I said.

I'm not a very good saxophone player, but it relaxes me. And a jam, no matter how cacophonous, is one of the few things that can make an escaped goat follow you back to the corral. But my silvery tone makes my dogs howl for the dead.

"Perfect," she assured me. "The coffee's all set the night before. He just switches it on."

I set an unhealthfully early alarm, thinking, *I'm hanging with real farmers now, and right at harvesttime. Better be on time.*

Three verses of marginal jazz is a lot of sonic force with which to pierce the Colorado Front Range at dawn on a Thursday. The stars were still out, and a frosty, pre-sunrise tang was in the air. For a little while, I got into it, swaying a bit—I was *meaning* every note, you know what I'm saying? Imparting a touch of vibrato into my phrasing. Then something in the very back of my brain told me to get a little worried. It felt like I had been honking away for a long time to, well, a dark farmhouse. The hemp field behind it was still a black velvet silhouette too.

At about 6:37 a.m., the door flew open. There was a bumping sound and a colorful noun describing dog offspring, followed two long seconds later by the hulking figure of Althouse greeting me on the porch. He was clutching a toe and technically "wearing" a dangerously small, barely knotted white kimono with the Japanese character for enlightenment printed on it. To this day I am very glad for the distraction of that character.

My final few notes kind of trailed off, as though a plug had been pulled from a turntable. I gave my colleague a weak *not a bad alarm clock, eh?* look. Bill's first words weren't coherent in any language in which I am fluent. But the moment the coffee was in our bellies and the kimono blessedly replaced by Carhartts, the first place America's most knowledgeable organic hemp farmer took me was into the field.

If Iginia is the heart and soul of the FPS, Bill is farmer-in-chief. He has a connection with the plant kingdom the way Jane Goodall has one with chimps. Whenever I come upon him in the field, there's something in his stillness and posture that makes him look a lot like one of the plants.

It was good to see him again. When we met five years earlier, he was a legendary purveyor of organic strawberries and giant dahlias (we're talking dinner-plate size) at the Santa Fe, New Mexico, farmers market. He moved to Colorado the moment hemp was legalized there, and I used that economic exodus in testimony to get the Land of Enchantment to legalize.

We strolled out past my camper just after sunrise, blowing into our hands. Once we reached the emerald hemp rows, hand-shaped leaves waving to us in the breeze by the tens of thousands, I asked Bill what was the number one lesson of hemp farming that I might bring to my own planned hemp projects. These would begin the following spring, in 2016.

"You're doing it right now," he said immediately. "Be on the farm every day. Watch the plants closely. If they're asking for something, give it to them. And if you can't be walking the field every day, make sure someone you trust is. Because if you miss two days, even one day sometimes, you can easily lose a whole field."

Bill Althouse is not a guy who says things to hear the sound of his voice. Working on the remote set of the film *Natural Born Killers* in 1993, he endeared himself to Oliver

Stone by saying simply yes when the great director asked if Bill's motorcycle could make it to Santa Fe and back with a needed prop, on single tank of gas. To accomplish this 4-hour round trip, Bill outdistanced pursuing state troopers outside of Taos.

"In general, Bill is a guy who wants things done his way," said Yamie Lucero, a San Felipe Pueblo member who also fled New Mexico to grow hemp. At 27, he was an intern at the Fat Pig Society when I met him but is now the youngest co-op member. "I won't lie to you: It can get crazy sometimes. But here's the thing: He's usually right."

Having fled an abusive car dealer father in Texas at age 16, Bill is possessed of granite toughness that makes him one of the world's great get-it-done guys. I've seen it a dozen times, from dahlia-fertilizing techniques to hemp-processing modes: If he can think it, he can manifest it.

In 2015, no one knew organic hemp cultivation like Bill. If he said, "Be on the farm and watch the plants," that's what I was going to do from here on out. I watched so closely over those next three days of harvesting that I donated my expensive Maui Jim sunglasses to the soil.[9] I found them months later, on Iginia's face. I was happy to make the donation in exchange for Bill's priceless piece of initial information.

As a goat herder, I immediately thought Bill's advice had the ring of truth. If you don't milk the goats twice a day, you don't have milk. It's about presence in your own life, as opposed to rushing around between judo, grocery shopping, and soccer practice. The reason my family has difficulty taking vacations (or indeed getting to town for groceries) is because you can't just ask a neighbor to swing by and milk your goats at sunrise and sunset. You have to keep a loving eye on them. They only let gentle, patient people milk them.

That's what made the biblical Jacob such a kick-butt shepherd that he pissed off his father-in-law. You can't phone in agriculture gigs.

The longer I farm hemp, the truer Bill's words prove. If you are a farmer, be on the farm. Even if just for an hour or three. If that doesn't sound like a fundamental element of a hemp business plan to you, or if a day spent getting muddy while being dive-bombed by mosquitoes isn't your idea of a good time, then independent hemp farming might not be the profession for you. It'd be like a dentist with a tooth phobia, or a preschool teacher who's not really into kids.

Work gloves are a telling piece of equipment in the new regenerative economy. Folks who always have a pair handy tend to be my kind of people. That includes potential investors. I had mine with me in the Fat Pig Society hemp field, and Bill noticed.

"What are we looking for specifically?" I asked him 3 hours into that October harvest day. The temperature had jumped nearly 40 degrees, and we were sweating, having clipped maybe 75 pounds of densely flowering hemp branches each, hanging them on the fence line. That fall Rocky Mountain sun has a way of singeing the epidermis, which is fooled by the overnight chill into thinking it's already winter.

Bill had been squatting and staring out over the field for 4 long minutes once I started keeping track. I tried to follow his gaze, but what he was doing was more nuanced than a shepherd watching a flock. He was cataloging marginally visible details, the way a cat sees things we don't. My friend seemed deaf to any human voice, flies buzzing his ears, and his ringing cell phone. He just kept looking. I remember doing a few awkward yoga poses, waiting until he came out of his reverie.

"Here, it's water first," he said finally, taking hold of my arm and guiding us along a late-planted row that was thus a little earlier in its development than the areas we had been harvesting. "Touch the soil—to your second knuckle. Watch the leaves."

We continued doing this for 45 minutes, covering an acre. Then, my own water bottle empty and this fellow 20 years my senior never having taken a sip, Bill pointed to a dry drip line.

"Look at this," he said, stooping. "Clogged at the mainline. If we weren't here on a day this hot, we'd lose a hundred fifty pounds of flower. Lucky today is a watering day for this row, or I might've missed it."

The hemp row in question sported leaves that were ever so subtly beginning to droop. The soil itself still felt slightly moist when you got an inch and a half down, though that would've changed quickly if the weather held. This is not something you would've noticed if you checked your crop by remote cameras, relied on a quick glance, or hired a neighbor to pop by. This required touch. This required love.

But we were in time. We devoted a half hour to shutting off hoses, fixing a leak Bill found at a spigot (most of that time was spent finding the adjustable wrench), and replacing the section of clogged drip mainline. It felt good getting wet. So good that by sundown I had the sax out again, serenading the crop ("The Pink Panther Theme"). I couldn't be sure, not having interviewed either of them on this specific point, but I got the impression that the plants at sunset were more receptive to my jam than Bill had been at dawn.

While the city-block-long, recently dry row of fragrant, bushy CBD plants had visibly perked up, that might have had something to do with the drink they were enjoying in efficient little plinks from the drip line. I was tired as I honked. Real tired, not first world tired. Ten acres is a lot to walk in 1 day. One acre is a lot, when you're paying close attention to soil and leaves. Farming keeps you in shape, if you do it right.

Three days later, Bill, Iginia, and another original FPS member, John Long, took me to a nearby commercial kitchen to teach me how to process hemp flower via decarboxylation.

Decarboxylation is the most ancient way of processing cannabis/hemp: in essence, activating a plant's cannabinoids through heating. Technically, as the name implies, you're removing a carbon atom. It's what happens when you light a spliff or make a CBD cookie. Like many enterprises, the FPS folks have since moved on to another processing mode (cold ethanol) due to volume demands, while I've stuck with "decarbing" so far (see "Decarboxylation" in chapter 10, page 169).

It was an invaluable day in the evolution of what became my own product. I hope the kindness embedded in the FPS members' willingness to share their expertise comes back many times over. But I mention it here because, apron-clad in that kitchen, I immediately saw the payoff of Bill's insistence on being on the farm. Each time we heated, stirred, and jarred the FPS flowers for the co-op's coconut oil–infused product (called Free Hemp), I was acutely aware that I had handled these plants, even if only for a short time. I had touched, smelled, and snipped them, while giggling with friends. Seldom-used muscles were sore from a sort of squat thrust routine executed in 12-hour shifts for days.

Being on the farm allowed me to process the harvest with more awareness and care. I have little doubt it resulted in a better product. And I felt better too. As would soon become routine in my life, I emerged from these harvest days muddy and exhausted but perceptibly even more at peace with the general rightness of creation than I had entered them. Without thinking about it, I'd spent a few days meditating, clearing my thoughts, breathing.

I've never heard of anyone regretting a field day. When you find yourself out of range on a farm, away goes that bank account discrepancy you've meant to check. Every problem seems not just solvable but minor: You're growing food. Everything's going to be fine.

So don't even think about pawning that little bit of the process off (growing the actual hemp), imagining you have more important work to do on the computer or at the trade show booth. The farmer isn't your worker. The farmer is *you*, or at least you plus your partners. Learn to be suspicious of clean fingernails. The last thing you want to do is farm out the farmwork.

Redolent of heated trichomes, I maneuvered the camper out of Fort Collins that October evening forever aware that engaging at the farm is a vital part of the process for anyone in a hemp enterprise. It's why, seven months later, my youngest son celebrated his sixth birthday planting hemp with the rest of our family in Vermont.

CHAPTER THREE

Owning Your Seed

Vermont, 2018

Seeds have the power . . . to counter economic monopoly and to check the advance of conformity on all its many fronts.

—MICHAEL POLLAN, *Second Nature*

E ager as I am to get us acquiring strong seed genetics and then into the field to plant them, I figure we're wise to understand what's going on behind the genetics-policy scenes before we're armed with the knowledge about how to choose our seed.

The first thing to remember is that hemp plants want a little THC. When you try to take it out, it generally comes back or it weakens the plant. Stripping THC in the name of some bygone drug war would be like stripping your blood of hemoglobin because a vampire might get drunk on it.

The reason we define hemp as we do at the moment (0.3 percent THC or less by dry weight) is because of a 1976 paper by the Canadian researchers Ernest Small and Arthur Cronquist. Here's the operative phrasing that guides our industry to this day (but not for long): "We arbitrarily adopt a concentration of 0.3 percent Delta9-THC . . . in young, vigorous leaves of relatively mature plants as a guide to discriminating two classes of plant."[1] Adopting the 0.3 percent part of this impactful sentence, as much of the world has, created a newly separated branch of the cannabis plant called "hemp." It marked the first time that hemp was specifically defined as something separate from psychoactive cannabis.

Back in the "Hemp for Victory" days, when the USDA begged farmers to grow hemp for the World War II effort (some of which ended up as the parachute cordage that famously saved future president George H. W. Bush's life) THC wasn't even known: It was isolated in 1964. For all of human history through about 1976, cannabis had always been one plant.

This is a visceral issue for me, because for a minute I was worried that our 2018 Vermont crop had tested hot. (*Hot* refers to a crop that has, arguably, more THC than current hemp definitions allow.) I opened the email from our testing lab that fall and felt that realization you get when you catch the flu that's going around: *Oh, you're saying this applies to me too?*

Suddenly my team and I were not invincible. Suddenly we were deeply empathetic toward a large number of our colleagues. Because suddenly, we thought, we were among them. And that team included the state hemp program coordinator for Vermont's Agency of Agriculture, Food & Markets, Cary Giguere. Along with his daughter, Erin, and son-in-law, Colin, who produce a line of hemp products called Vermont Farmacy, we had become cultivation partners back in the spring of that year. When he heard that I was a free agent in the Green Mountain State with the dissolution of my original Vermont partnership, Cary set up a meeting at a Killington pub and pitched, "We've got organic-ready family land. You've got a promising cultivar. Just saying."

Now, back in New Mexico after eight months of hard work together culminating in a successful Vermont harvest, I had just come inside the ranch house from dashing around the Funky Butte arroyos—Julie Andrews had gotten into the grapes, then hopped a fence. That's when I saw the message that contained our Vermont peak flower results. At first I thought the report contained a decimal point typo. I texted my partners, "They must have meant '0.27 percent,' not '2.7 percent,' right?"

The same seed, that same season, tested under 0.3 percent in Oregon. And in previous years it didn't test hot in official tests in any of the four states where I had grown or provided it. But it turned out I wasn't reading the results wrong. I had conducted the wrong kind of test.

These were peak *flower* results. This is important. Peak flower samples are those clipped from the top three inches of the highest (or "cola") bud on cannabis plants, where the plant is bursting with cannabinoid-soaked trichomes. In Vermont, draft rules proposed that farmers be allowed to sample from the more farmer-friendly leaf of the plant. (Leaf testing is friendlier because the leaf contains cannabinoids, but in lower levels than the flower does, so a leaf-tested crop is less likely to go hot.) Regardless, for a split second I felt like Calvin bringing home a report card he's not eager for his parents to sign.

Dang! This was the first crop I had cultivated that had received USDA organic certification. Our group celebrated with freshly harvested hemp waffles drenched in home-tapped maple syrup. We saw it as a huge step both practically (the three-acre harvest became at least 15 percent more valuable) and philosophically (one's chosen crop going from a Schedule I federal narcotics felony to USDA organic in five short years—not too shabby). Now, suddenly, I was concerned that I might not be able to make my share of the harvest into Hemp in Hemp.

Concern that one's potentially lucrative crop might have to be composted makes life perhaps a bit too exciting. I recall a few anxiety dreams when that flower result first came back. Fortunately for all Vermont farmers, for more than a year Cary Giguere had been researching hemp cannabinoid testing. Vermont didn't yet have an official protocol, but its policy makers were working to develop one in anticipation of the pending full legalization of hemp. Since long before we teamed up, Cary had been developing an industry-supporting THC-testing protocol whose early proposed versions included, among other testing options, sampling of "young, vigorous leaves of relatively mature plants."

Interestingly, this friendly leaf-testing idea came from the same 1976 paper in which Small and Cronquist arbitrarily (their word) picked 0.3 percent THC as the definition of this new hemp branch of the cannabis family tree. In fact, leaf testing is recommended in the same sentence of the same paper.

I've always imagined Small and Cronquist collaborating in their Ontario labs, reasoning that if you're going to one day cause thousands of farmers sleepless nights worrying needlessly about THC that won't show up in their final product, you might as well choose a testing mode that doesn't stack the deck. But when I emailed him in 2019 to ask about his motivation, Dr. Small, who is now a member of the Order of Canada, the nation's second-highest honor, said that the 0.3 percent THC level "was based on previous analyses of THC content in thousands of plants [with testing of] upper, younger leaves." In an earlier interview he further clarified, "It was calculated to be the concentration that naturally best separated the two groups."[2] By "two groups," Small meant "hemp" and "psychoactive cannabis."

Regardless of the research methodology, the fact is, the 1976 paper that landed at 0.3 percent also suggested leaf testing. By developing a plan that would include this unimpeachable source (and implementing it as an a priori policy for the 2018 season in advance of final regulations), Cary saved dozens of Vermonters' hemp seasons four decades after the paper's publication. That included ours, but he had no reason to think it would. Leaf testing is just a fair way to test for THC.

In 2019, Cary also proposed an additional testing-method option based on a genetic sample taken right at germination that can determine whether or not the plant is hemp. Many farmers are fans of this mode because they'll know early in the season that their crop has passed its THC test. I prefer leaf testing later in the season in part because I'm not sure how regenerative such genetic testing modes are, and in part because I believe it is easier for a borderline plant to pass a leaf test.

Here's proof of how good policy really works: Remember that 2.7 percent flower THC result? Our third-party leaf test on the same crop came back at 0.3 percent. Clean as a Bernie Sanders conscience.

This is what the crop deserved. Even a flower test of 4 or 5 percent is not going to prove a commercially marketable option for a psychoactive black marketer. Dispensary ganja flower starts around 12 or 15 percent and goes to 30 percent or above. Our crop was obviously hemp, and a fair test confirmed it.

In 2017, 30 percent of Colorado and Kentucky hemp farmers tested hot. That's because of biology: Cannabis produces cannabinoids for predator defense, pleasing humans, and the other reasons we've discussed.

From a policy perspective, when one-third of farmers aren't allowed to harvest their crop, it's a neon red flag that THC-testing protocol needs an upgrade. The best upgrade would be not to test farmers' crops at all. As of

2019, depending on a crop's level of hotness and a given state's policy, it can be and often is destroyed because of policy minutiae best described not just as pointless but as insane. While some states allow mildly hot crops (up to 1 percent THC) to be retested, used for compost, seed, and fiber, or even cannabinoid extraction via cannabis dispensaries, other states' policies still mandate crop destruction and even threaten prosecution.

Making THC levels irrelevant to farmers across the land is one of the two baseline policy goals that will ensure that this time, as a new major agriculture-based industry emerges, the farmers are the folks who primarily benefit. The other policy goal is called the "genetic level playing field": meaning the right of farmers to own, develop, and replant their seeds and clones. This happens to be both a human right and a key economic pillar. No corporation will ever restrict my family's ability to plant food. From a business standpoint, it's also the key to a diverse and robust hemp industry controlled by independent farmers.

At this moment in cannabis-human relations, if you want to grow hemp in the United States, your first move (once you've started cultivating mycelia and collecting livestock dung) is to acquire a cultivation permit in your state. These cost anywhere from $25 to $1,500 per season. My New Mexico permit sets the farmer back about $800, depending on acreage and the number of cultivars, which is pro jargon for "hemp varieties."

One day hemp won't require any special permitting. If we rally with sufficient ferocity, it'll be one day soon. But even with the passage of the 2018 Farm Bill, federal law mandates that each state submit its hemp program plan to the USDA for approval.[3] Then each farmer (and in some cases, also each processor and seed provider) must have a permit. So call it "legalization" if you like, but these are not the same rules that farmers face for growing, say, tomatoes. We'll get there.

It's up to each state to establish its regulatory framework—this is the rulemaking process that follows the passage of a state's hemp law.[4] A genetic level playing field and THC irrelevance are the two essential components that set up independent hemp farmer-entrepreneurs for success with a supportive regulatory climate. Here's why.

The Genetic Level Playing Field

As many hemp-program administrators already recognize, farmers must have access to all hemp genetics for at least five years before any certification process is established. This means that hempseed (or clones, tissue culture, pollen—however plant genetics can be transferred) from any state that meets federal hemp definitions may be transported and cultivated, without the government deciding for farmers which varieties are kosher. Having the option to own and develop these genetics is a nonnegotiable baseline for a farmer-based hemp economy.

Farmers must control their genetics because some hempseed providers want to be the hemp Monsanto or Syngenta. Surprise, surprise. They'd like to sell you their seed every year, meaning you can't replant it. These would-be monopolists are banking that farmers have been cowed by the past half century's neo-serfdom nonsense. Nope. As a sole option for farmers to acquire seed stock, that model is a nonstarter. Generally speaking, the independent entrepreneur is wise to look instead for open-source seed providers who don't aspire to be genetics moguls.

"You can tell because they'll offer you some kind of a 'right of replication' form when they provide the seed," Wild Bill said. "It's how we do it." That's also how I roll with my genetics.

You'll have to pay a price for open-source seed—generally more up front than non-open-source genetics—but then you own it. You can replant it, eat it, make it into a nail-care product—do what you like with it. The only exception is that you have to pay a royalty if you decide to market genetics deriving from seed provided.

Now, I like to think I'm Mr. Reasonable Middle Ground. (Who doesn't?) So let's say you're a seed developer who argues, "I've spent a decade and hundreds of thousands of dollars working on a cultivar that I like because of its [purported] THC stability, high CBD level, or the protein content in its seed: Why should I let farmers buy it once and then replant it themselves?"

First off, because that's how agriculture has always worked since the end of feudalism until recently. Out on the prairie, Pa Ingalls was not enslaved to the wheat company from which he bought his seed. Seed saving is part of the traditional farming process. But sure, if you want to sell your genetics cheaper than an open-source provider does with the stipulation that farmers must buy it again next year, fine. As long as there are plenty of alternative options from which a farmer can choose. A level playing field, in other words. There are scores of terrific open-source domestic seed varieties available to farmers, whether you're growing for seed, fiber, cannabinoids, or soil building. Once the industry has matured a bit, these can all compete for seed certification in a state or region.

Say a farmer is in it for the long haul and doesn't want to buy seed every year. Old-style seed developers would like to maintain the 20th-century process for becoming a provider of a non-open-source variety. Under this model, the farmer starts with proprietary breeder seed and works for several years through the established stages (foundational, registered, and certified seed) until he is essentially the representative of a particular seed company in a region. That's also fine, if that's your choice.

As a farmer, while the buy-seed-every-year model is not exactly my definition of food security, it's one option that might work for some enterprises. Let's say your product is delicious hemp hearts (dehulled seed). Everyone in my family eats this omega-balanced superfood in our morning goat yogurt. It could conceivably be cheaper for an entrepreneur to buy and grow the same seeds every year from the same provider. And, arguably, it could result in more year-to-year consistency than having to build your own stable seed stock. As long as your provider stays in business and maintains supply, quality, and cost.

For most independent entrepreneurs, though, owning genetics will be the wiser choice. And the key word is *choice*. A level-playing-field policy means both models (open source and non) are allowed. And it's easy to implement: All a state genetics policy has to say is that permit holders' crops must test within federal hemp THC definitions. It's really that simple.

I recall shivering under my blankets at a high-altitude Colorado campsite in early April of 2018 when my frosty, one-bar cell phone pinged. It was Cary Giguere, in Vermont.

"I'm in Montpelier at the Senate chambers right now, working on the new hemp bill," his note said. "What is the most farmer-friendly, federally compliant wording for acquiring seeds or clones?"

I blinked, sat up, and felt around for the coffee thermos. This was real-world legislation that could affect thousands of people, and millions of pounds of carbon sequestration. In a very uncharacteristic move for me, I kept it short and sweet. In about 20 numb-fingered seconds I typed:

> Registrants may purchase or import hemp genetics from any state that complies with the federal requirements for the cultivation of industrial hemp.

I am proud to say that this is how Vermont law reads today.

It's easy to demonstrate why a level playing field is so vital: The states that implement strong genetics policies and the few remaining that don't are as different as night and day. The proof is in the numbers. In Oregon, for example, which has a hands-off genetics policy, 7,808 acres were cultivated in 2018, third most in the nation. (More than 1,300 farmers are projected to plant 46,219 acres in the Beaver State in 2019, a sixfold increase.)[5] Across the Columbia River in Washington, by contrast, farmers suffered from a legacy hemp law that mandated offshore importation of seed that in most cases they couldn't replant or even grow for flower applications. Have a guess on 2018 acreage? One hundred forty-two. And 105 of those were the Colville Tribe's project. (Washington's policy, and its number of farmers, improved markedly for the 2019 season.)

Here in New Mexico, our policy makers saw what the successful hemp states were doing and listened to what we initial Land of Enchantment farmers wanted, especially with genetics. As a result, we have a terrific program that has a per capita record-setting 353 applicants cultivating more than 2,000 acres in its first season. Quiet, humble New Mexico might even find herself among the national hemp leaders before too long. Plus, 2,000 acres means a lot of carbon sequestration.

Right now, Vermont's hemp law and regulations are best-in-nation; in addition to the genetic level playing field mandated in the state's hemp law, Vermont is leading the way toward making our second vital policy (THC irrelevance until retail level) standard operating procedure. Vermont's regulators want hemp to succeed, according to 48-year-old Giguere, because, "our dairy industry has cratered, and farmers need alternatives."

As a result, Vermont has a hemp program designed for independent farmers to thrive. Cary's motivation to cultivate hemp while a state employee was almost identical to mine to cultivate hemp while a journalist: He had to plant. In my case, the wildfire that nearly made my family climate refugees and cost us much of our goat herd was the catalyst.

Cary came to it from a higher altitude. "The rebirth of the hemp industry is too important a phenomenon in human history to ignore," he said when he suggested we team up. His boss, Anson Tebbets, Vermont's secretary of Agriculture, Food & Markets, agrees: Six months after Cary and I decided to cultivate together, Secretary Tebbets showed up (in loafers and Oxford shirt) to help us plant. Dude got dirty.

There's an old hair-restoral commercial where the hirsute founder assures us, "I'm not only the Hair Club president, but I'm also a client." That's how Cary, who also heads the state's pesticide-management division, operates. The curly-maned gentle giant gets up early, heads to work in Montpelier, and with almost poetic legislative wording, empowers his state's farmers. Then he comes home, farms a bit himself, and finally enjoys a microbrew beside the pond with his wife, kids, and often me. Then, as all farmers do, he gets up and does it again.

The Giguere clan is also, like perhaps 96 percent of Vermonters, a family of sugar makers. Culturally, Vermonters make maple syrup like San Franciscans design websites. My favorite is the Giguere family's Grade A Very Dark syrup, the porter of the maple world. The point is, Cary understands the plight of the small farmer, because he is one.

"We're a state of independent farmers—the mountains don't allow for many huge fields," Giguere said. "It makes sense for policy to allow them to thrive. Obviously farmers should choose their genetics."

Policy translates into the quality of the hemp in the ground, which is what really matters. The reason for that is best expressed by Michael Pollan in *The Botany of Desire*: "American cannabis farmers are the best gardeners of my generation." Which is to say, domestic genetics are often superior to, and are certainly competitive with, the non-open-source varieties currently deriving largely from overseas. Folks who have been growing underground for a half century have, in the realest way, been conducting agronomic efficiency studies. They are hemp's Gregor Mendels. So if hemp is helping someone you love, hug a former black market cannabis farmer.

As with any crop, farmers know what they want to grow and how to grow it. If they want to encourage a new property in a cultivar—say, more CBC in their crop's flower—they'll breed for it. I'm doing such cannabinoid fine-tuning this year. Limiting this flexibility for any reason, let alone for the benefit of wannabe neo-Monsantos, would be like a regulator telling Steve Jobs in 1981, "Laptop and watch development is okay, tablet and cell phones no, plus you can only use chips made abroad that we approve." We, the farmers, will make our own "chips," thank you very much.

Another key issue that a simple genetics clause like Vermont's resolves is the "certification" red herring. The way a state seed certification protocol generally works, seed specialists in the agriculture department, usually in sync with a coalition of certifiers known as AOSCA (Association of Official Seed Certifying Agencies), decide what varieties of a crop are on the approved list for commercial production. This, in general, is a good process. We are in such an early developmental phase of our industry, though, that very few domestically "certified" seeds even yet exist.

But this isn't a problem. Again channeling Michael Pollan's assertion, our open-source domestic cultivars will populate a strong portion of the future certified-seed family across the land. That's precisely because homespun genetics are often outperforming European and Canadian "certified" imports in a number of areas, including both seed production and flower composition. I've seen it in the ground repeatedly.

Don't take my word for it. Just witness the recent efforts of Canadian seed companies to purchase US-developed cannabinoid genetics, including those of my Oregon partners, to see this reality in play. In other words, you (the new independent hemp farmer) benefit from the genetic level playing field because you have the opportunity to demonstrate that you are competitive with any other farmers. As American hemp farmers did in the 18th and 19th centuries, you and I will build our seed stock.

Since any genetic-quality claims are anecdotal in this early phase of the industry, I'll add that I think it's safe to say, by any agronomic standard, the more genetics farmers have at their easy disposal, the better. Edison, remember, tried 6,000 potential lightbulb filaments before landing on tungsten as the solution.

Ahead of USDA policy announcements, a few states already got rolling on hemp genetic certification, and more may be doing so by the time you are reading this book. Colorado, for one, has been doing it in an appropriate way: If you meet certain benchmarks for germination and THC levels in each of the state's bioregions, you're in. And so far one US cultivar has qualified. Which is far better than none.

As several hemp-program coordinators have pointed out to me, establishing seed certification protocol has been a requisite step for many commercial crops. Because our industry is in its infancy, though, we, the farmers, are the ones who should be deciding exactly what we want to grow, and how we want to develop our genetics. And most of all, we demand to own our seed. We have consulted with Wendell Berry. We've been tipped off about how the game has been rigged for half a century. As the young industry matures, the rules must be different for hemp genetics.

Given that hemp spent 77 years in the penalty box, we all have to start somewhere when it comes to developing the ideal genetics for our microclimates and end products. If you're like my hemp professor Edgar Winters, this is a lifelong project. I rarely visit Oregon without either Edgar or Margaret handing me a tiny watercolor brush and asking me to paint grains of pollen on a promising flower in the greenhouse. Hemp, I'm learning with every crop, is amazingly adaptive—the second crop in a region is commonly more robust than the first, often regardless of a cultivar's origin. And some cultivars are more adaptive than others.

If a state ag program does decide it wants, after 5 or, better yet, 10 years, to institute a seed certification program, the level-playing-field launch period will have served its purpose. Farmers—both independent entrepreneurs and commercial seed-development entities—will have had a window in which to see what grows well in their ecosystem. Not enough time to really develop a breeding program, but enough to pass germination and related tests. What we're *not* going to allow is what happened to farmers of corn, soy, and wheat (and alfalfa and on and on). Business as usual—namely, trapping famers in a debt cycle to seed and herbicide companies—is not part of the hemp brand. Lest this pride in domestic cultivars come across as jingoistic, I'd like to add that I absolutely love that farmers all over the world are developing terrific regional cultivars. *Vive la différence!* We want Champagne and Chardonnay, Parmigiano-Reggiano and Roquefort. And when farmers from Ohio to India build their seed, I and many others are hopeful, that will begin to stem the farmer suicide crisis.

The excellent program admins in states like New Mexico and Vermont realize that hemp will not be ready for certification until farmers have had years to work with as wide a variety of genetic options as possible, to determine what works best in each region. Even Colorado waited five years from the implementation of its first-in-the-nation hemp program before launching its certification protocol.

Is there a downside to allowing the genetic level playing field? I'd offer two caveats: The first is to be careful from whom you get your genetics. "Many farmers got junk genetics this first year," Brad Lewis of the New Mexico Department of Agriculture told me. That bummed me out to hear because I and others had good options. I got a lot of last-minute panicky calls from farmers, but, oh well, there's always next year.

The second is that some states allow you to choose your genetics, but you're on the hook in a potentially menacing way if they test super hot. Still, choosing your own is totally worth it. Especially compared with the restricted genetics nightmare my colleagues at the Colville tribal project have been through. As one of the few brave entities willing to forge ahead under Washington's abysmal initial hemp program, the project suffered a double insult: The European seed the tribe planted in 2017 and 2018 was nothing like the quality of the domestic cultivars I planted in my personal projects. And the tribe couldn't replant the imported genetics even if it had wanted to.

Still, project coordinator Richter was shrewd: She figured the tribe would be ahead of the learning curve when a real hemp program was instituted. And that's how it's playing out; under the new federal hemp law, tribes are treated like states. Recently, Richter sent me a draft tribal program to read. I replied with a set of comments that basically came down to, "Be like Vermont."

Vermont's slam-dunk law and promising pending rules help explain why it has the highest hemp acreage per overall farm acreage in the nation: 1,820 acres in 2018, with, tellingly for our indie-boosting purposes, all but two applicants cultivating 10 acres or less. It also explains why I've cultivated there (and in Oregon) for three years, until my own state came online: These have been the most friendly hemp business climates.

As a starting point for ensuring a level playing field in your backyard, make sure your state's hemp program doesn't mention genetics at all, other than to say that they must meet federal hemp definitions. If you see the words *approved cultivars* in your state's law or regulations, it's code for "indenturing the independent farmer to seed companies." Don't rest until you have this kind of wording struck. From there our mission is to restore the very definition of *hemp* to its historical one, where THC is a nonissue to the farmer. That's the day when cultivating hemp is treated no differently from cultivating tomatoes.

THC Irrelevance Until the Retail Level

This, our second vital policy endgame, means that permit holders in any state can grow any variety of cannabis they like, regardless of potential THC. The burden of worrying about minute variations in a commercial plant's chemical composition will be removed from the farmer. THC will matter only if a final product exceeds locally determined levels for adult-use regulation—the way alcohol and tobacco are regulated. Otherwise, it's nobody's business what variety you grow, any more than your variety of tomato. And yes, this means that at the federal level, the definition of *hemp*, as opposed to cannabis, will go away. All cannabis will be, as it has properly been for most of history, regarded as one plant.

When I first discussed federal THC irrelevance with national cannabis and hemp lobbyists five years ago, they thought I was kidding. *Reefer Madness* was still too recent. Commercial hemp wasn't yet legal. Now they're telling me it might happen within 10 years. I think it'll be sooner. In 2016, when I walked into the office of Thomas Massie, congressman from Kentucky, a conservative Republican hemp supporter who represents the dry-on-Sunday Christian County, the first question he asked me was, "Do we need to raise the THC definition of hemp to one percent?"

"That's a good start," I replied. "Then three percent, then no federal THC definitions at all." (Switzerland and Tasmania already have 1 percent hemp definitions.[6])

Regardless of the timeline, the law of the land must have the feds out of the THC-regulation game entirely. States can set their own THC levels, but only for retail-level products. The farmer will be free to grow whatever she likes without interference.

While we work to implement this policy everywhere, farmers' hemp crops will still have to be tested for THC. So an immediate stop-gap measure is to ensure that THC testing is conducted from samples taken from the leaf of the plant, rather than from the flower. As we discussed, a leaf-tested crop is less likely to go hot. This will be combined with farmer-friendly THC-calculation formulas. These measures maximize farmer-entrepreneur opportunities under current law.

The reason we must stop punishing farmers for these microfluctuations in THC levels is this: A hemp plant harvested in the field very rarely goes directly to a customer. The fun little secret in the early hemp industry is that processors don't need to worry about THC. Only farmers do.

Unless you're processing your hemp flower via decarboxylation, you are almost always going *way* over 0.3 percent THC at some point in the course of your processing journey. For instance, if you process your hemp flower in pursuit of a CBD product in a cold ethanol extractor, it initially comes out as a concentrate, what hemp professionals call *crude*.[7] This crude might contain, say, 80 percent CBD and 7 percent THC. Although still much milder than dispensary cannabis, that's 23 times higher than current hemp THC limits.

Standard procedure at this point is to dilute the brown pasty mixture (with coconut oil, for example) until it is back below 0.3 percent THC in the final product, or to run it through further equipment that removes the THC. No one worries about this—not entrepreneurs, not regulators, not buyers at chain stores. And they shouldn't, because the final product in the store meets federal hemp definitions. So why, during cultivation, do we obsess over small THC variations in flowers?

Since a field test usually has no connection to the final product, it's totally unnecessary. It's like testing beer when the final product is orange juice. And it has devastated some of my friends' enterprises.

To give one example, the family-owned Salt Creek Hemp operation on Colorado's Western Slope could have benefited from a sensible testing policy like Vermont's proposed regulations. "A fairer testing protocol would have saved our 2017 crop," Salt Creek's Aaron Rydell told me recently when we were trying to rustle some of his cows near Grand Mesa

National Forest. "I got my first wrinkles dealing with destroying that 0.34 percent crop we had doted on for eight months. And the test results were within the margin of error for the machinery that the state used."

These Salt Creek peeps are some of the most down-to-earth, shirt-off-their-back, willingly taxpaying operators you could ever hope to meet. They gladly made every effort to follow all of Colorado's regulations. For instance, the genetics they acquired that year had a certificate of analysis (COA), proving that they had passed the previous season's THC test. Their crop tested hot partly because a flower-testing regimen punishes farmers for being too good at their job. A hot test really means a healthy crop.

THC, remember, is something the plant wants. So (and this is the really horrible part), a common method for avoiding hot crops is to tweak nutrients (like nitrogen) and, when necessary, harvest early. In other words, be a worse farmer to meet admittedly arbitrary rules. This is not a viable long-term industry game plan. But today you actually have farmers trying to impede their crop's robustness.

I know my own seed best, so I'll speak to the Samurai cultivar I've been developing. All the copious seed and fiber it produces contain negligible THC, regardless of what the flower test shows. Just to be sure, Cary Giguere and I tested the seed from our 2018 Vermont harvest. No detectable THC.

In addition to testing protocols promoting bad farming, they aren't even standardized; the equipment might be calibrated differently from one state to another (or even in different parts of the same state).[8] Plus, as Ernest Small himself has pointed out, THC levels in a plant can vary significantly between morning and evening.[9] All of this adds up to a policy that needs to change, posthaste.

Cary is on it. He's a soft-spoken visionary. Sometimes when he runs a policy idea by me, it takes weeks for its shrewd impact to sink in. But the results to date have always been the same: His policy decisions have a way of palpably lowering my blood pressure. I mean this quite literally; if you'd strapped me to an EKG for the entire 2018 hemp season, you'd be able to chart each solid policy decision on the readout.

After proposing leaf-testing and early plant-genetics sampling as options in Vermont's draft regulations, Cary went one better: He proposed that Vermont not require THC testing for dioecious cultivars at all. Only sinsemilla crops would be tested. If you're marketing, say, hemp hearts, hemp plastic, or phytoremediative services, why should your flower's THC matter? It shouldn't. These are seed and fiber applications.

If you do grow sinsemilla-style for flower applications in Vermont, under the draft regulations, once a test is passed, Vermont asks no more THC questions until the retail level. The future is here in New England: THC is irrelevant. And that is how to launch a hemp program.

A friendly THC-testing formula is almost as important as leaf testing. Cary improved my cardiovascular health for a third time after the season, when the stove fires were lit, the waffles eaten, and four feet of snow blanketed our recently harvested field. That was when he sent me the innovative formula he had in mind for ensuring that Vermont's hemp program will fit into whatever USDA testing guidelines emerge.

Based on policy-maker chatter in 2018, it appeared that the good folks at the USDA were likely to implement a cockamamie testing standard called "combined," or "theoretical" THC. That means THC in both its nonpsychoactive "acid" (THCA) and "active" (delta-9-THC) forms would figure into a crop's THC test. All hemp/cannabis plants have some of both types at field-testing time. Such a standard is misguided because the acid form of THC should make no difference to anyone. THC is not psychoactive in its acid form. The drug war is over. Cannabis won. The day has come to relax about nonactive THC, as a society.

But if you've got to calculate theoretical THC, Cary recognized, then the best strategy was to make Vermont's formula as farmer friendly as possible. You know, so people who are following the law, like Salt Creek Hemp, can actually harvest their crops.

Cary ensured Vermont was among the first to implement what might now, thankfully, become a standard: He reasoned that the feds could accept a total theoretical level of up to 1 percent THC, as long as the delta-9 level was at or below 0.3 percent. And this is the formula he decided to work with:

$$\text{Total theoretical THC} = (\ [\text{Delta-9-THC}] + (\ [\text{THCA}] \times 0.877)\ )$$

Here's why this formula is so fantastic: If your leaf test comes back with 0.3 percent delta-9-THC and 0.8 percent THCA, you are golden—you end up with 0.96 percent total theoretical THC, and 0.3 percent delta-9-THC. You pass the test and your hemp, as it should be, is considered hemp.

It's a sensible formula. Even though it's never happened, when hemp was legalized, some bankers and farm insurers expressed concern that a hemp crop could be "abused" by imaginary psychoactive cultivators who for some reason might want to sneak ganja by as hemp. Cary's formula prevents that. Secondly, the helpful "multiply by 0.877" part of the formula (when you multiply by a number less than 1 you get a smaller number than the one with which you started) is derived from established research into the percentage of THCA that is likely to be converted to the delta-9 form during decarboxylation.

"I learned about it from a German paper," he explained.[10] "That 0.877 constant is now widely accepted as accurate." When this formula winds up becoming the nationwide standard, we'll have hemp administrators like Cary Giguere to thank for the resulting surge in the independent hemp industry.

Plus, it's flexible. "If the USDA decides that a 0.9 percent or a 5 percent total theoretical THC is the cap," Cary said, "we can modify the variables to meet the standard. I think of good policy as a moving target. We'll make it work for the farmers."

———

Outside of policy, there's one ultimate reason why we independent hemp farmer-entrepreneurs are going to win these two important battles. I mean besides the fact that we're in the right, we want it more, and we'll fight to the death to win. That reason is bioavailability. Many in the industry firmly believe that hemp performs better with some THC in it.

Now, some folks who would like to be your seed salesmen are touting that they've developed "patentable zero THC" varieties of hemp. Hemp is a big tent, and if you want Diet Coke just because there was a drug war last century, depriving your plant of a component with which it has evolved for millennia, go for it. But folks like Margaret Flewellen of Natural Good Medicines are of the mind that whole-plant products are superior when they contain, well, the whole plant. As she puts it, "I and our customers believe the most effective products come from full-flower sources, rather than from isolating individual chemicals in the plant as though we're making a pharmaceutical."

And the belief that hemp containing some THC is superior extends beyond flower-based edible product. When THC is a nonissue, hemp fiber can achieve ideal performance as well. Just ask my closest colleague, Margaret's husband. The white-bearded, perpetually energetic Edgar Winters has grown hemp since 1957, making him probably the longest-cultivating hemp farmer in North America. One time as we sat in our midseason field at sunset, I asked him how he got started.

"It was in Alabama, when I was seven," Edgar told me 60 years later, still sounding as if he left Alabama last week in the way he unveiled every rounded vowel with great patience. "My Choctaw granddaddy Joseph Pate saw that the family cotton farm had what you'd call a competitive advantage by continuing to use hemp baling twine, rather than that plastic bullshit that was coming online in the postwar cotton industry. So we kept growing hemp long after prohibition started, for the fiber." That plastic baling twine continues to pollute and disfigure the agricultural landscape to this day, including mine, by the millions of tons. Salt Creek's Joe Koeller told me his crew pulled "probably a hundred pounds" of plastic from their remote field, mostly baling twine.

Edgar maintains that hemp with some THC in it contributed to the strong reputation that his family's fiber enjoyed. "Nobody knew what THC was per se," he said with a wink. "My guess is ours had three, maybe four percent in the flower."

Ganja farmers have reason to support THC irrelevance too. Coming from what is for the moment called the "cannabis" side, a good example of THC irrelevance in action can be seen in the initial venture of my state's former governor Gary Johnson—once a Republican, now a Libertarian. His first cannabis product was a mint aimed at seniors with arthritis. It contained very little THC—practically "hemp" levels. The marketing thrust, as he explained it to me, was that the mint had enough THC to help the anti-inflammatory process but not enough to get Grandma too high.

This is where the cannabis/hemp industry is headed: one plant, with innumerable applications. On the one hand, folks will be cultivating industrial cannabis with high (say, 18 percent) THC flower that never sees market; the retail product will be a high-quality fiber or nutritious seed, for instance. On the other hand, some enterprises will cultivate cannabis with low THC (say, 0.1 percent, as in Margaret's Zenith hemparettes) for actual retail use of flower. And we'll see folks cultivating at all THC levels in between—whatever works best for their product. Might be a post-workout ointment, might be a space capsule door panel. Regardless, the essential point is that it will be the farmer's choice.

Oregon recognizes that short-lived definitions are going away; in 2019 the state legislature passed a law that allows interstate and international shipping of psychoactive cannabis. And yet we're still arguing over minuscule THC levels in a farmer's hemp? That ship has sailed.

Because we independents make up the largest body of hemp players at the onset of the industry, our choices and voices (including those of you who leap in now) are, accordingly, strong.

That means we're going to win this. But it's a corollary of Margaret's Law (hemp is a year-round endeavor) that, as if you didn't have enough work to do in the field, you're also now a political operative. Prepare to spend at least three or four sunny mornings stuck in marble hallways talking to your reps or policy-making bureaucrats. We're playing by a new set of rules this time around, and you and I are helping write them.

These days when I discuss policy with legislators, I feel like one of those finger-wagging free enterprise libertarians telling the government that my industry knows best. I'm always reminding our public servants, "We're bringing in a billion-dollar industry to your state." And guess what? They listen. It's happened so often in so many states over the past few years that I've stopped being shocked. *Boom*, New Mexico institutes a fair genetics policy. *Wham*, Washington repeals an out-of-date planting restriction. *Zang*, South Carolina allows retesting of THC samples up to 1 percent.

I think one reason hemp regulators are listening is that everyone who glances at the tail end of the monoculture-farm-era economy can see that healthy, regional agriculture manifests a new form of patriotism. Regulators see the declining production numbers in struggling soil. If we don't reboot farming in a regenerative way, it's a genuine threat to national security. Hemp, accordingly, was declared by executive order to be one of the nation's "essential agricultural products" that should be stocked for defense preparedness purposes.[11] This happened in 1994, 20 years before initial legalization.

The endgame for those of us who find the delineations between hemp and cannabis irrelevant is a return to how this plant has been treated for 8,000 years: as a valuable multipurpose crop whose cultivation must be maximally encouraged and minimally regulated. In other words, we're doing this gold rush differently. If there's an element of initial federal cannabis policy or a state hemp regulation that we don't like, we'll have it changed.

And we might have to. As I was checking a few policy facts with Cary the other day, he told me I had reached him in DC, where he had been discussing USDA policy all day. "Even when I handed them Small and Cronquist's study, some congressional staffers laughed at our leaf-testing plans for Vermont," he said. "They might or might not make it into the final regs."

They won't be laughing for long. I suggested to Cary that anytime someone's initial reaction to a regulatory protocol we farmers propose is, "But that's never how the agriculture system has been regulated before," a good reply is, "How's that been working out for the planet's farmers and soil?"

If a state plan that includes our essential policies is in danger of being rejected by the USDA in its initial plans, it just means our public servants haven't heard from nearly enough independent farmers yet. We'll fight until we win. This time the farmers are in charge. That's the only way to rebuild soil and rural America.

CHAPTER FOUR

Wild West Genetics

Now, let's face it, chum. I'm not leaving till I sell you something! Now, what'll it be? Just name it!

—DAFFY DUCK, *The Stupor Salesman* (1948)

With soil prepared and permit acquired, now we're ready to sketch the genetics options for the farmer-entrepreneur. What should you be looking for as you source your seed? The answer is deceptively simple: Grow for what you want to harvest. Cherry tomato seeds are different from heirloom Roma seeds. Similarly, there is a significant difference between genetics intended for just CBD (and other cannabinoids residing in the flower) and genetics intended for nearly any other purpose. Regardless, advises master breeder Edgar Winters, "look for someone who has provided good genetics in the past, who has a COA and is a permitted farmer in his state."

To that I would add: Examine the cultivar's cannabinoid test results and germination rate carefully. (The latter should be north of 85 percent.) If the seller can't provide both of these forms (COA with cannabinoid test results and germination rate results), I wouldn't buy the seed. I also suggest visiting the source farm, so you can actually look at seeds, plants, and the overall operation. Is it clean? Organic?

Third, shop early: November is much better than March. By spring, everyone is busy, many breeders have run out of seed, prices are usually higher, and few providers have time for the many questions you're wise to ask. And finally, ask for the kind of "right of replication" form that we discussed in chapter 3. Own your genetics.

Some states issue specific seed-provider permits in addition to cultivation permits. (Hey, bureaucratic budgets need holiday bonuses too.) If the state in which you're sourcing seeds has such a permit, you might ask to see your putative provider's.

Genetic sourcing, especially CBD genetic sourcing, is a massive caveat emptor situation as of this writing. Not for me, since I am my seed provider or I know my provider personally. But nearly everyone who comes to me asking about genetics, CBD genetics in particular, is a refugee from a sketchy seed situation. In other words, I get a fair number of emails that start with, "Help! I just got burned on bunk seed."

Here's one that just came in from the Empire State:

> Hello Doug,
>
> We are searching for some hemp seed and thought you would be a good person to reach out to. We grow organic hemp along with other crops in upstate New York. We have a bag of seed that we just purchased and it came loaded with Indian meal moth and it looks like the bag is mostly shot.
>
> Kind Regards,
>
> KM
>
> Head Grower

But now you've done your homework, you've found your reputable provider, and you're on a farm visit. How do you buy in an informed way? Start by examining the seed closely. This is not a hard-and-fast rule, but generally speaking, CBD seeds tend to be smaller and seeds for other purposes to be fatter. But although I've seen some monster seeds, even most of the bigger varieties aren't much larger than a BB.

Edgar seeks larger, bulbous seeds in varieties that are intended for food. "Look at these fat ones, bursting with oil and meaty hearts," he told me not too long ago as we pored over a Spanish cultivar in his greenhouse.

There's considerable variety in how the walnut-shaped hemp shells look on the outside; some are tiger-striped and some a solid cream color. I've harvested seeds that were downright purple. But a robust seed generally cannot be cracked between your fingertips. At least it should be very difficult. So on your farm visit, lay down a five-dollar bill and ask the breeder if you can squeeze a few.

Once you've decided what kind of harvest you're aiming for and you're ready to buy, it's time to start thinking about how much to plant. Our "tri-crop" dioecious Samurai cultivar (tri-crop = the Big Three of seed, flower, and fiber, plus, when applicable for a project, phytoremediation) has been planted in seven states as of 2019, at several different planting densities. I'd say we get best results from about 15 pounds of seed per acre, more if fiber is the primary application. (Higher-density planting produces tighter, straighter plants, which are easier to process for fiber applications.)

We've also planted as low as eight pounds per acre, which maximizes seed and flower production by giving the plants the room to branch out and reach for the sun. But this wider plant spacing (15-inch bases between plants, center to center) means the crop is more difficult to harvest via combine. That's because rather than plants, you get hulking 10-foot trees like something out of *Little Shop of Horrors*.

Those looking to grow for CBD (or other exclusively flower-based applications), have several options: mixed-gender seed, feminized seed, and clones. These all have one shared goal in common: an all-female crop. Her flower is your harvest. Though I welcome fertilization of the flowers in my product, most farmers at this peak gold rush moment don't. In 2018, four out of five US hemp farmers were trying to maximize one cannabinoid: CBD. Fertilization of the female flower can lower that CBD level, and thus the value of a wholesale crop. So if you're going the flower-only route and start with seed, you can either purchase seed for which you pull males, or shell out for purportedly feminized seeds.[1]

Flower cultivators can also annually purchase clones, which are genetically identical clippings from mother plants.[2] Or they can buy mother plants, from whose branches they can repeatedly clip their own clones. They stick the baby clippings, or "plugs," into some form of soil or growing medium (many folks use Rockwool) until their roots establish and they grow a few inches. If you want to use goop that purports to stimulate root growth, please be sure it's OMRI-compliant, so your crop can qualify for organic certification.[3]

When the plugs have stabilized, they can be transferred to the field. Starting seeds or clones in greenhouses allows outdoor farmers in cooler climates to add a few extra weeks to the growing season by transferring already-sprouted seedlings when the last frost has passed. But a frequent rookie faux pas is transferring the plants immediately outside without acclimating them for a few days with increased outside exposure each day. Rounding out your starting-point options, "tissue culture" is a high-tech variety of cloning done under sterile conditions that is having something of a moment as of this writing. None of these cloning options are my jam, personally. But I see why some folks choose the cloning route. One advantage is you come away with large volumes of similarly sized females, all producing the high CBD that the wholesale market is seeking at the moment. That makes automated transplanting easier. One disadvantage is that the plants are arguably less hardy, especially over time.

Edgar is passionate on the topic. "Cloning is not how these plants work in nature," he said. "And they don't work well for very many generations —the mother plants get weaker over time. So if you go with clones, you gotta source new mother plants periodically. Myself, if I get hold of strong genetics from clones, I pollinate 'em and turn 'em back into seed-producing plants."

Even though practically everyone I know who has cultivated from clones has had at least some major issues with post-transplanting die-off or greenhouse pest infestation, one thing you have to say in clones' favor: They are efficient. And if you invest in enough mother plants (which can cost $100 or more each), you can soon personalize your clone operation. Ideally in a solar-powered greenhouse using real soil and sunlight as much as possible.

"We had to move to clones for the quantity of product that we produce," Bill Althouse told me. "The uniformity, the ease of planting, the regularity of the planting cycles, all made cloning the best option. We can turn hundreds of thousands of soil-ready starts around on a three-month cycle in the greenhouse."

When a new generation of clones is ready (roughly three inches tall) and it's the right season for outdoor planting in Colorado, the Fat Pig Society's farmers can just, in Bill's words, "plug these little sisters in the ground, two thousand an hour—*boom boom boom*," care of machines called transplanters that will sink entire flats of baby hemp plants right from your tractor, at spacing that you choose.

It's fun to be part of this kind of planting: Two farmers ride on the transplanter in mounted chairs behind the slow-moving tractor, feeding plugs into the rotating planting slots on the transplanters. This is the same kind of equipment that strawberry farmers use. A high-end Italian model called the Baby Trium runs about $25,000. It even has ergonomic chairs for the always-sore farmer.

It's at harvesttime that uniform plants can really make a larger-acreage farmer's life easier. "Early on we had plants of all kinds of sizes in the field, which complicated harvesting at scale," Bill said. "Now we've got a replicable system."

Bish Enterprises, an innovative Nebraska company, engineers and markets superefficient flower-harvesting tractor attachments, which neatly slice and swing a whole row of flowering plants up a conveyor belt and into a catchment bin. I've watched its Hemp Handler 6031 model bring in three-quarters of an acre of flower in an hour.

"I think that the time you save makes it worth the [$37,000] investment," said Bish's 37-year-old president, Andrew Bish.

Even with the clone advantage of uniformity, if I were growing a female-only crop, I would still grow from seed. And I would not pay the extra thousand or two per pound for feminized seed. I've already disclosed that I'm a dioecious guy, even for my flower crops. It's a hormonal-balance theory. Sinsemilla crops are, when you get down to it, sexually frustrated females. All those sticky trichomes they produce are part of a desperate attempt to snag a stray grain of pollen in order to make babies. My motto is "everyone's happier when they're dating" and so I use seeded flower in my product. That is sacrilege to today's wholesale processors. But I'm not aiming for maximum CBD per flower. I'm aiming for an ideal cannabinoid–terpene ratio—the entourage effect.

But even for flower-only farmers who aspire to an all-female crop, feminized seed can present problems. Cannabis not only wants its THC, it also wants to be dioecious. Maleness creeps back in. This has already resulted in lawsuits in Oregon from folks who paid top dollar for feminized seeds, believing all their seeds would be usable, only to find a few of those pesky boys in the mix.

This issue of "stability" in seeds is a complex one. CBD farmers crave it. Understandably, they want the same crop every season. Knowledgeable growers speak of "F1" and "Bx1" generations of seed, referring to the number of generations of breeding that have gone into that seed.

Here again, my philosophy is somewhat against the grain, so to speak: I look for diversity in my genetics. I want each season to be a distinct vintage. Plus, I seek oddballs—what breeders call *unicorns*. These are individual plants with unusual traits, some of which you can see (phenotypic) and some of which you have to test for (genotypic). It might be an individual that is fast germinating, frost hardy, high in protein, or in possession of distinct fiber color or unusual cannabinoid content. Here we see another advantage of listening to Bill and watching your crop every day: You're likely to find a unicorn you'd like to select and breed. *Vive la différence* once again.

Even for farmers devoted to growing a uniform sinsemilla crop, feminized seed can be a crapshoot. You still are going to have to examine every plant to make sure she's a female, so why not pay less for seed that doesn't claim to be feminized? And if you're a CBD farmer wanting to bring only females to maturity, one of the items on your daily farm walk checklist in the early part of the season can and should be, "check for males." You have plenty of time to identify and remove them before they mature.

The science of seed feminization is kind of cool, though. I've visited farms that employ a variety of techniques to achieve it. At Functional Remedies in Colorado, for instance, farm manager Hunter Konchan demonstrated a method of adding silver to the plant's feeding, which evidently scares away the Y chromosome.

"The silver, introduced at the right time in the plant's life cycle, reduces the levels of the hormone ethylene, leading to female offspring," Konchan told me as we toured the vast Functional Remedies greenhouses. Konchan, by the way, is proof that farming is one of the digital age's hottest professions: He was hired before he graduated ag school at Colorado State.

Regardless of his genetics choice, when a sinsemilla farmer is visualizing his plant spacing in the field, he wants to provide plants room to branch and flower. I've found flower-only crops are best grown like psychoactive cannabis, meaning they should be given much wider spacing than our tri-crop applications demand (30- or even 60-inch bases) and much fewer seeds per acre—around three pounds. Which is a good thing for your budget, because CBD seeds are much more expensive than grain, fiber, and tri-crop seeds. Some folks grow in greenhouses, which are measured by square foot rather than by acre. So to figure out how many plants to cultivate in a greenhouse, measure out five-foot bases until you hit a wall. Although greenhouses that utilize natural light and real, local soil can be great, I myself am primarily an outdoor-cultivating farmer. I believe outdoor cultivation results in the highest-quality hemp, and definitely the most carbon sequestration.

Now on to the big question of peak gold rush pricing: Just what can the first-time hemp farmer expect to pay for seed? The first thing to know is that over these next few pages you will almost certainly encounter outdated numbers. Oh, how I hope so. We are in the Wild West phase of cannabis genetics pricing. Because of young, seasonal, unstable markets, prices can swing hugely both over the course of a year and from region to region. That last part always surprises me a little. One could always hop on a plane. Furthermore, folks offer all kinds of different pricing structures. In gold rush terms, the seed merchants are the shovel sellers. They get paid whether or not you grow a successful crop. When you first explore the seed-market landscape, you feel like a prospector in *The Treasure of the Sierra Madre* with these hyperinflated genetics prices and colorful characters making exotic offers to you from dodgy websites.

But here are some rough numbers: For CBD seed, you might pay between $3,000 and $6,000 per pound, depending on the CBD levels in the COA for the cultivar. Feminized seed can cost up to $7,000 a pound. Clones range from $3 to $10 per plug, depending on quantity and region. Keep in mind, as we discussed, that you're planting far fewer seeds/clones for CBD crops (around 3 pounds per acre) than you are for seed/fiber/tri-crop harvests (15 pounds, give or take).

For a different perspective on flower genetics, I'm going to outline the business model of Rich Becks, of Chimney Rock Farms in Colorado. I do this as a countercurrent to my own philosophies, since Rich is a guy I respect, and is very professional in his breeding. He does not agree with my total right of replication outlook or my aversion to feminized seed.

"It costs me fifty K to develop a new line of genetics," he told me. "Someone who wants to buy a clone for four bucks and make an infinite supply is basically free-riding off our work and investment."

So Chimney Rock offers a "farmer membership" that costs $5,000 per year. It includes online consulting and access to Chimney Rock's standard operating procedure (SOP). Becks supplies feminized seeds to each member at what he described as a below-market price—enough for five acres.

"These are high-performance genetics that are developed for full-spectrum products with specific cannabinoid ratios and terpenes," he told me. "My job is to make the member successful."

And since feminized seed can wind up costing north of a dollar per seed, the Chimney Rock model can make sense to farmers looking to have reliable genetics plus consulting delivered to their door. Knowledge can indeed be priceless, especially the first time out of the gate. Like a first-time parent, a new hemp farmer is going to have a hundred questions over the course of her debut season. I'd go so far as to say that the most common mistake first timers make is not investing time and resources in gleaning knowledge from more experienced farmers.

Becks favors feminized seed since it's a labor-intensive and technical process that in his view makes each seed worth the price. "That's because you add another twenty-five to fifty cents in soil, trays, heat, light, and labor and you end up at half the cost of a clone with a much better plant thanks to its taproot."

When it comes to replicating Chimney Rock genetics, Becks has a two-tier system. Farmer members can clone Chimney Rock genetics for themselves but not sell them. But Becks said "grower partner members" are licensed to produce Chimney Rock genetics in their state.

"I supply the foundation genetics, maintain consistency, do the R and D and marketing," he said. "I send them leads and they just grow and pay me a royalty."

So that's CBD genetics in the time of Wild West pricing. Now on to seed prices for my favorite hemp apps: everything else.

For cultivars intended for seed/fiber/tri-crop harvests, you might get price quotes as low as $2 per pound for staid old stock that you can't replant, or up to $2,000 a pound for proven open-source cultivars that you own upon purchasing. If you do the math, after five seasons of owning your genetics, even that higher price pays for itself. For the past couple of seasons, generic seed/fiber/tri-crop cultivars with right of replication tend to land somewhere in the middle of those two numbers: They usually run between $30 and $100 per pound. Then you own the seed forever.

In five years, the genetics landscape, worldwide, will be radically altered from its current form. Prohibition will be a fading memory, supply will be much larger, hemp regulations in most states will reflect the level playing field we've been mapping, and legit state certification programs will begin to sprout. As well, the USDA and the AOSCA seed-certification agencies we discussed earlier will begin weighing in with guidelines.

I hope that in a generation, readers of this part of the book can have a deep (if sympathetic) laugh at the situation their parents faced. Your established, major crop, grown in the millions of acres, kids, was once a brand-new gold rush, full of speculators and shady operators, wearing the same Carhartt coveralls (or lawyer ties) as the people trying to save humanity.

I'll endeavor to keep folks abreast of the situation via social media, live events, television, and short-form social media in coming years. A book is a magnificent thing, but it is static. And that's one of the few things that hemp genetic prices (and markets) aren't.

That said, I'm glad we're documenting the Wild West phase. It's a legitimate part of hemp history. Plus, if I were entering the hemp world now, I'd be more concerned about genetic legitimacy than prices. That's because today's prices are not necessarily so prohibitively expensive as they seem, if you look at the current market conditions for the harvested crops.

It's all well and good (and no small feat) to grow a pretty crop. But if you're paying so much to get your seed, what is your potential return? We'll start with flower markets.

Even at $4,000 a pound for seed, a CBD variety can potentially make a legitimate upper-middle-class living for a 20-acre farmer. Especially for a clever, righteous, farm-to-table provider who is willing to do the extra work and take on the added risk of marketing a value-added product —often called a *secondary market product* or a *consumer packaged good* (CPG) product.

Because there are so many value-added products on which a farmer-entrepreneur might focus, outlining the metrics for success can be tricky. My five-year plan with Hemp in Hemp is to scale up from current, tiny 1,000-bottle runs to 10,000 annual top-shelf three-ounce units that wholesale at $50. And that outlook involves utilizing both the seed and flower parts of the plant. Plus I own my genetics, so that budget line item is zero. But if Hemp in Hemp meets that year-five production goal and all 10,000 units sell, that will mean a gross of $500,000 on minimal acreage. Might the production runs expand from there? Maybe. If folks demand more than 10,000 units, then scaling up while maintaining farm-to-table quality will be a good problem to have.

But say you simply want to sell your seed, fiber, CBD flower, or processed crude on the open market. I hesitate to get too deeply into that kind of plan, both because of Wendell Berry's warning about falling victim to the vicissitudes of fluctuating commodity markets and because wholesale market prices are also on the roller-coaster ride of the Wild West phase. They could crash at any time—I think we have about three years until markets start to mature and stabilize.

But since many folks are going to follow the CBD herd anyway, let's do the agronomic math: If you're planting for CBD, you use those roughly three pounds of seeds per acre. (There are about 27,000 seeds in a pound.) Meaning you invest about $12,000 per acre in seeds. The New Mexico Department of Agriculture folks estimate a $6,000 per acre cost to cultivate hemp, outside of genetics—for soil preparation, drip irrigation lines, and so forth. So let's call your total cost of cultivation $18,000 per acre. You have a great year, and you harvest 1,000 pounds of flower per acre on your 20-acre farm.

The low end of CBD wholesale raw-flower (often called *biomass*) prices over the past couple of years has been about $20 per pound. (It sometimes goes as high as $1,000, and averages $150-ish.) But at $20, you'd clear $2,000 per acre, or $40,000 on your 20 acres. Not terrible. In a $100-per-pound wholesale market, that same harvest would net you $82,000 per acre.

Even at these most conservative calculations, a hemp crop can yield four times what a corn, soy, or wheat crop does. But if you turn those 20 acres into the 10,000 units of value-added product for which I'm aiming in the medium term, and successfully market the full run (no fait accompli and involving much more sweat equity than simply selling wholesale flower), your gross could be as much as $320,000 before marketing, packaging, delivery, and other post-harvest costs.

Marketing your own product is a gut check, but that route has the highest long-term earning potential for the independent farmer or co-op. If you scale up to 100,000 units, you'll be well into seven figures of annual revenue. Plus you'll be providing a great product to customers and sequestering thousands of tons of carbon. So feel free to start rubbing your palms together and making greedy "hoo hoo *hoo*!" noises. As long as you do it right and have a multi-season game plan, both your bottom line and the planet can benefit.

Another option for the flower-only farmer, as long as recent wholesale prices hold, is to "toll process" your flower into crude. That means using someone else's processing equipment to turn your flower into the raw materials of your product. This is what Salt Creek Hemp does with its harvest: It has a reliable toll processor for its crude, and a capsule-making company turns the crude into the final value-added product.

On the wholesale side with crude, your 1,000 pounds of flower that tests at 10 percent CBD will give you about 90 grams of crude containing 80 percent CBD per pound of flower. (That's 20 percent efficiency in processing, meaning you lose 80 percent of the flower mass that you harvested.) Wholesale CBD crude prices hovered around eight dollars per gram in 2018. So now, instead of $20 or $100 for raw flower, your pound of processed flower is worth $720, less your processing costs.

If you're thinking of going with a toll processor, interview several, and pick someone with a track record of not gouging farmers. One reason we didn't sell our Vermont harvest as wholesale crude in 2016 is that the dispensary with which we were negotiating wanted half our harvest in exchange. That was a crop we lovingly worked on for 240 days. They would have run it through their equipment in 1 day.

If you've got the up-front capital, you might consider becoming a toll processor yourself. Then you can process both for yourself and others. This is what Dexter Rice of Sub-Zero Extracts in northeastern Colorado does: He invested $175,000 for a medium-scale cold ethanol processing rig and related facility build-out costs. Now he extracts cannabinoids both for his own Nature's Love line of products and for other enterprises that want to lease his 120-pounds-per-day equipment for their own products.

Six figures might sound expensive to a bootstrapper trying to save the family's Midwest farm via hemp. But Dexter took the leap and it's paying off: When you generate $250,000 in combined product and toll processing revenues annually, a $175,000 investment quickly pays for itself.

Or you can scale up your processing equipment in start-up-friendly stages, as Margaret Flewellen of Natural Good Medicines has done. She started with a $7,000 ethanol extraction machine from Eden Labs in Seattle, the smallest one they offered. "Our equipment has grown as our business has grown," she said. "We're on our third unit now."

Those in the flower game are wise to always keep in mind that CBD is just one cannabinoid. As a whole-plant advocate for maximizing bioavailability and the entourage effect of artisan cultivars, I'm not a fan of isolating specific elements of the cannabis plant. But in the name of thoroughness, I won't completely ignore these markets. The equipment exists to isolate just about any cannabinoid or terpene you like. And the markets are real, if volatile.

When I walked into the green room for some coffee before a hemp conference in 2018, a former real estate agent–cum–hemp broker was touching up her makeup and speaking loudly into her phone. She was half lip gloss, half hustle. This is what I couldn't have avoided hearing her say, even if I had wanted to: "Oh, sure, Yuri, two kilos is as easy as one.

We'll send them both right off, no extra shipping. Yep. Should be there by Thursday."

When she hung up, she hollered to her partner across the room, "The Ukrainians want two keys now!"

At this point, I jumped in. "I have to ask . . . was that CBD isolate?"

"CBN," she said. "Can't keep it in stock."

"And, um, if you don't mind my—"

"Fifteen thousand dollars a key."

I can pinpoint this as the moment I realized that the hemp industry is for real, worldwide in its fungible markets, and about to become very, very big. Specifically, it was my first of many reminders that there are 110 other cannabinoids in play. Today I got an email asking me what I charge for CBG (cannabigerol) isolate. Heard of that one? One peer-reviewed study listed "anti-inflammatory" and "antiemetic" as two of CBG's qualities, and another found it inhibited colon cancer cells.[4] Vermont farmer Rye Matthews of Northeast Hemp grows in part for CBG and told me he finds it "uplifting and analgesic."

Folks growing for organic seed applications (like hempseed oil and hemp hearts, or value-added products using them), rather than flowers and their embedded cannabinoids, can expect to harvest 1,000 pounds of seed per acre. That's the new normal, up from the 600-to-800-pound Canadian average of just a few years ago.

In the wholesale market, organic seed for food is worth between one and two dollars per pound. If you paid $300 per acre for your planting seed ($20 per pound), then your profit, for much less work than a flower crop, is $700 per acre wholesale (if you received one dollar per pound for your harvest). And here is where seed ownership is so crucial: The following year, your seed is free. Plus your own seed is available whether or not seed companies come and go.

Dioecious crops are considerably less labor intensive than sinsemilla crops because you're not obsessing over gender and flower manicuring. If you toll-process your seed harvest into a hemp-heart product that you market yourself, again, you're talking much higher revenue than wholesaling whole seed. Organic hemp hearts retail at around $15 per pound.

Owning seed-processing equipment is a less expensive proposition than owning flower-extraction rigs. A high-end, single-screw, seed-oil press will set you back about $15,000. Capacity varies, but you'll be able to press 250 to 500 pounds of seed with a small press in a 10-hour workday. So if you're expanding, you might need more presses, pretty quickly. Hemp Oil Canada, then the world's largest seed-pressing company, was on its fourth equipment upgrade in 10 years when I visited its Manitoba facilities in 2013. And that was before US legalization.

Whether you're buying seed presses, dehullers, or bottling machinery, today's players all emphasize the importance of high-end equipment, whether new or used. You get what you pay for, and top-of-the-line equipment companies tend to offer better customer service. That's essential for a nonmechanical guy like me.

Chad Rosen, president of Victory Hemp Foods, a hempseed processor in several US states warns, "Nothing ever comes ready out of the box. So finding a company that still exists when you call is a wise investment. We learned that one the hard way."

Rosen said that a reliable US seed-oil press manufacturer is Wisconsin-based AgOilPress. In 2016 in Vermont, our group used a Swedish Täby press, which worked slowly but terrifically.

Fiber is, as of this writing, financially viable only for those with large acreage. That is, unless you are a craftsperson hand-making the hempen equivalent of a Stradivarius. Which is not a hypothetical: two companies are already offering hemp-body guitars: Silver Mountain Hemp Guitars and BugOut Guitars. Small-volume, ultra-high-end hemp-based consumer goods are a viable small-acreage market for fiber.

Generally speaking, though, you need about 3,000 acres to feed even a small fiber-processing facility. But that makes it an ideal market for a cooperative, as we'll discuss. The fiber is there in the field regardless of your primary harvest application. So if 25 big-state farmers team up to process the conservative 6,000 tons that 3,000 acres will provide, there's significant value there.

Just how much value depends on how entrepreneurial this hypothetical farmer co-op is. Baled "mixed fiber" drew perhaps $200 per ton in 2018 (a bit more from buyers under contract with the few existing North American fiber facilities). But say you form a co-op that separates the hurd (the inner core of the hemp stalk, sometimes called *shiv*) from the bast (the long outer fiber, which makes such fine clothing, paper, plastic composites, and next-generation battery components). Then you have multiple streams of secondary-market products.

Hurd is the hot fiber app at the moment. Marty Phipps of Old Dominion Hemp asks $15 for a bagged, 33-pound bale, which is used as horse bedding. Hurd is also popular for the fast-growing, carbon-sequestering hemp construction market (known generically as hempcrete building). Spill cleanup and moisture absorption are additional up-and-coming markets for hurd.

That relatively simple process of separating the bast from the (in this case) desired hurd through a technique called decortication quadruples the fiber harvest's value, to $800 per ton. All you have to do is open your gate once per season for the co-op fiber-collection truck, and you share in the value-added revenue of the bagged hurd's final price. There was so much demand for Phipps's product that he still, as of 2018, had to import some of his hurd.

Now, someone has to buy, operate, and maintain the decortication machinery (which ranges from $20,000 for a small-scale hammer mill to $8 million for true industrial-scale facilities), plus the bagging operation and climate-controlled storage. Here is where the Farm Bill comes in particularly handy: Hemp farmers now have access to agriculture grants and loans.

Keep in mind that when you do decide to market any value-added product, you bear the burden of elements like quality control, insurance, and payroll.

"You're on the hook for liability once you decide to be the entrepreneur," Roger Gussiaas of Healthy Oilseeds in North Dakota reminded me.

But value-added product will often be the better option with any part of the hemp plant for the enterprise with a long-term game plan based on top-shelf quality. If you stake your enterprise on current gold rush wholesale prices, you're probably in for a shock. A wild initial pricing ride is a normal phase in the growth cycle for any booming new industry—not that knowing this makes the bite any easier for the new farmer-entrepreneur who makes the wrong decisions.

The best way to insulate your small-acreage enterprise from the coming wholesale roller coaster is to develop your own product line.

But I hope that in so doing you remember Margaret's Law: You might want to call off any plans for downtime for a half decade or so. The uptime is generally a lot of fun, though, when things are going well and you get to skip through fragrant hemp fields for a living. Plus you get to save humanity.

CHAPTER FIVE

Adventures in Planting-Gear Malfunction

Oregon, 2018

*You realize that the sequence of preparatory activities is
so long you will never get to the intended task. So you
go fishing instead.*

—Patrick McManus,
The Night the Bear Ate Goombaw

The easiest part of hemp planting is figuring out your seed
depth, plant spacing, and watering protocol. The hardest part
of hemp planting is getting your farm equipment to implement those instructions.

In fact, I'll tell you right here to plant at a half-inch depth in moist
soil that allows for good seed-to-soil contact and thus maximum germination. Doing that with the 7-to-15-inch spacing we discussed will
occupy 47 minutes of your 20-hour planting day. The other 19 hours
and 13 minutes will mostly be spent under a terrible device called a seed
drill. By, say, 11:00 a.m., generally the emotional nadir of a planting day,
you'll be dirty, bloody, very hungry, and thinking, *Huh, I would've thought
my first hemp-planting day would involve more actual planting of hemp.*
By lunch you should consider yourself in very good shape if you're even
sinking the first seeds in the ground. In case it helps you remember that

you're not alone, this diary of my group's three-acre 2018 planting of the dioecious Samurai cultivar in Oregon's Emerald Triangle reflects how planting day usually goes.

7:05 a.m.: Survey of Field, Yoga, Return to Child Mind. The ideal date range for sowing hemp is a latitude-factored-on-climate-change issue. It'll vary from late March to mid-June depending on your spring weather forecast and cultivar. In 2018, it is at the end of May for our field above the Rogue River. By this point we've cultivated billions of microbial communities before the seed even hits soil—mostly by leaving it alone for 20 years.

Not long after sunrise I set my coffee on the tree stump that marks our snack stockpile and tool dump near the gate to the field. After a few Sun Salutations, the whole thing looks so doable. I'm sure we'll have our 50 pounds of seed in the soil in no time and I'll be tubing the river by midafternoon.

I should know better. By 2018, I am aware—as I wake in the farmhouse of my mentors and partners Edgar and Margaret up in the hills of southern Oregon's famous cannabis-cultivation region—that before noon we'll have basked in two dozen nerve-curdling delays. This is not my first hemp rodeo. I've chased goats, woodchucks, and one determined family of wild pigs out of hemp fields.

After a baker's dozen plantings, I have learned that the only certainty will be joys and hassles we can't dream up. For instance, the Pacific Northwest version of the—ho hum—Anthropocene epoch's annual millennial wildfires won't start for a few weeks in Oregon, and they will last for more than five weeks. But as always, I am willfully forgetting the coming realities of planting day. Spring has sprung. So right off the bat, I'd probably be happy in the DMV.

Being outside sets up a struggle between logic and endorphins, between deadlines and love, where the right brain wins every time. As you stretch, you're smelling forsythia and raspberry blossoms. Working in the dirt. Your office has no walls. Courting hawks land in nearby limbs. Nothing else exists. For those unused to the feeling I'm describing, it's called sanity.

From a practical perspective, this "child mind" is what makes you forget last season's planting nightmares. It is probably some chemical wafting out of healthy soil that casts an indisputable spell of forgetting.

This is, really, the essential component of childhood—you don't know, or don't care, what's coming next.

It's not only last year's seed drill delays that you forget. Your product's bottle caps don't quite fit the bottles? Your state's regulators are sticking with the absurd "field out of view from road" requirements for another season? Whatever, that was yesterday. Today is planting day. The ultimate now.

7:19 a.m.: Return to Barn for First Human Error–Caused Tractor Breakdown. The wise farmer approaches planting day very much the way a pro ballplayer approaches spring training. It's intended to get the cobwebs out. But Major League Baseball is smart enough to have 37 days of practice games. We farmers have to wake up, get dressed, and immediately pour lubricants into the wrong reservoirs in tractors.

Terrible sounds and smells alert the group to the problem. In 2018, our perpetrator (not mentioning names, he is just playing an assigned role) avoids eye contact by checking irrelevant tanks with a dipstick. Then the tractor expires into a profound quiet. Our planting day stops before it starts.

This, of course, happens when the temperature is still frosty, and the last thing anyone wants to be doing is unscrewing metal plugs. The next 27 minutes are spent draining one disgusting fluid, pouring in a second, and remembering that we meant to run to town yesterday to pick up a third.

7:46 a.m.: Talking Big. This important phase of planting day commences when, already three-quarters of an hour behind schedule and clustered around the stalled tractor and seed drill, your whole team is now on-site. Just seeing a bag of hempseed unleashes passion. The infectious excitement about the season opening in front of you all results in conversation that goes something like this:

"We can probably do two hundred fifty thousand units," your partner gushes, pouring a bit of test seed into the seed drill reservoir from a 25-pound bag balanced precariously on his shoulder. "These babies look like they're ready for it."

Before you can decipher that remark, the tractor-fluid situation gets straightened out and the engine turns over, leading to a group cheer. The ice is broken.

The aged diesel motor is loud. You shout louder. The hawks scatter. You and your team continue crunching numbers, visualizing the killing the enterprise is going to make when this superlative crop finds itself on shelves.

"Gonna be a great season," you agree, ignoring the fact that implementing your colleague's 250,000-unit suggestion would mean 25 times the storage you have dialed in for the flower harvest alone.

As the seed drill is attached to the tractor in a sort of awkward Iwo Jima re-creation, you spend some moments wondering if they award prizes for Most Righteous Farmer of the Year. Before getting a seed in the ground, you tend to put the cart before the oxen.

In the business cycle, planting time represents what you might call the R and D retreat, or the spitballing phase. Some good ideas do come from these field meetings. But really what unfolds represents the primate love of daydreaming. It's pleasant to visualize that "lying on the beach with an umbrella drink" moment that provides the final scene in 73 percent of movies produced in the 1980s. Everything is ahead of you.

7:51 a.m.: Tractor Moves. Leading a parade of choking farmers and dogs, the farm conveyance crawls 200 yards to the field, churning roughly Bhutan's annual petroleum output. This is one reason my product labels boast of a petroleum-free harvest. The planting, usually but not always, has been a different story.

8:04 a.m.: First Seed Drill Malfunction. There comes a moment on planting day when the final distractions fade. You feel an all-systems-go sensation. You've built soil, acquired your genetics, and prepped your field. Your seed has germinated at 95 percent in the 100-seed paper towel test you conducted as soon as you brought it home six weeks earlier. The tractor has bumped its way to the east side of the field, something that seemed wildly improbable half an hour ago. There you plan to make your first "pass," which is farmer-speak for the bundle of rows you plant each time your tractor does a lap.

Something clicks. The whole crew feels it. An internal timer signals that you've daydreamed long enough. Between fast-moving foggy hints of rain and skin-singeing teasers of how hot the day may get, everyone shoots one another an effervescent thumbs-up or shaka. *Let's get to work.*

This, according to the universal calendar of hemp, is when the seed drill fails. As the walls of our bubble of forgetting explode around us again on May 28, Edgar and I shoot each other a glance that says, *Oh, right. This.*

This is my fourth year of planting delays. His 62nd. We know our day has changed. We will have to spend many gory hours resolving this kind of SNAFU.

The seed drill (also known as a *grain drill*) is a device invented to punish us for something (maybe for staying still and farming at all, rather than wandering around seminomadically after caribou, wildebeest, and bison, the way we're hardwired to do). It's a nonmotorized machine hitched to the back of a tractor (or oxen team), basically a storage container with carefully calculated leaks that drop seeds down a series of chutes from the bin to the ground as often and as deeply as you calibrate it to do. Theoretically.

Like the tractor itself, it's supposed to make agricultural endeavors easier by improving on the time it would take actual human beings to plant seeds. Instead, working with a seed drill is easily the most maddening element of planting season. Not the only maddening element. Just the most reliably maddening. More practically, seed drill–maintenance delays ensure that agriculture remains about as efficient as it was on the first planting day along the Euphrates.

We appear to be trapped in a constant here, which I call Fine's Law of Abandoning Traditional Economic Rituals, or FLOATER. This constant establishes that in mechanized agriculture (defined as farming that employs machines rather than hands, hand tools, or livestock), a mission critical problem with a poorly designed, factory-made piece of crap will occur exactly once per pass during the first morning of a given year's planting season.

It can vary, but early in the day when everyone and everything is rusty, the time it takes to plant a pass plus deal with the malfunction leading up to it usually totals around an hour and a half. We have about 60 passes in front of us this day.

For a long while all the hawks can see are eight booted feet protruding, midfield, from under a tractor and its seed drill attachment. All they can hear is the occasional expletive when yet another socket wrench attachment proves to be just the wrong size.

Despite the delay, spirits are high in the long-angled morning light. That's because the mood in the field is that of a home birth. We are hemp midwives, and loving it. If you speak to most midwives, they'll tell you it's a pretty joyful occupation. A perpetual birthday party. And in our bodies as we plant any crop, oxytocin is exchanged as in any parent-child relationship.

Plus as a farming group, enough of us know that the pace tends to pick up in the afternoon. Even during the worst moments of FLOATER despair, it helps to keep in mind that the hemp will get planted. It'll just take 10 times longer than you've budgeted.

I haven't yet heard anyone say, "Dang, planting day was just too much of a pain in the ass. I decided not to go through with it." I have indeed heard such a sentiment following harvest quagmires. But not at planting.

The brain is a remarkably flexible chemistry lab. It can secrete, at electromagnetic speed, any emotion for which the situation calls. The sequence of planting day emotions is: Bliss. Frustration. Elation. Repeat. Unless you really do plant a small-acreage crop by hand, though (not a bad idea), just don't imagine for a second that you're immune from the FLOATER constant.

In our case at 8:04 a.m., what we notice before a single seed has dropped, before we have even "calibrated" the rate of planting on the device, is that the seeds aren't dropping at all. Something is clogged.

During the brief periods when a seed drill is operational, the hopper at the top of the machine, filled with thousands of seeds, drops exactly one (yeah, right) down each opening in a line of such inch-wide openings that extend across the bottom of the hopper. Each opening has a roughly seed-sized exit that is connected to one of those long chutes. The chutes feed a drive wheel where the seed emerges near the ground. And each opening plants one row. That's why a seed drill is supposed to be faster than hand planting: It plants multiple rows simultaneously. This drive wheel is what you'll calibrate to plant seed at your desired spacing.

Probably the biggest problem with seed drills today is the composition and design of the chutes themselves. In principle, they're just four-to-eight-foot-long straws, like elephant trunks. The primary issue is, for the past half century, they have usually and inexplicably been made from a variety of fake rubber that invariably cracks mid-tube when exposed to sunlight but fuses chemically to the two areas where the chute gaskets attach: at the hopper (the top of the chute) and near the ground (the bottom of the chute).

This makes it all but impossible to remove a chute from its metal fastening pins when cleaning or maintenance is needed. No reason it would occur to the designers of every seed drill in the world that a farmer might need to access the most essential part of the device—the one that actually plants the seeds.

On this now gloriously humid late-May morning in 2018, the issue is what you might call "be kind, rewind." The farmers who previously leased our seed drill (I'm guessing sometime in 1973) didn't clean it out at the end of their planting. Probably too blissed out by oxytocin. And the local farm-supply guys didn't bother to check.

As a result, an invisible-to-us glob of moldy seed cholesterol is stuck in the least accessible part of 6 of our 10 chutes. This is dangerous because there is no way to know what kind of seed formed the clog. Pro farming is an industry where crop contamination is a major issue. Along with our hemp, we could be planting soy, lavender, or opium. Also, the fake-rubber seventh chute is, predictably, disintegrating, spilling out seeds indiscriminately.

No seed drill repair task is a good one. Either you crush your knuckles trying to access key areas of the device, or you engage in extended duels with a family of recluse spiders in the barn while looking for spare parts. I can save you some time on this one: The part you need resides under an unmarked bucket of old fertilizer sitting atop a pack rat nest. It's another constant.

One moment you're in your midwifery bliss mode, and the next your colleagues are trapped under a sketchily jacked-up half-ton piece of machinery, bathed in WD-40 while consulting a manual written in Swahili. In between testy requests to "pass the dang adjustable wrench—no the other one—the one that locks," your crew starts making macabre jokes like, "How many hemp farmers does it take for automation to make life easier?"

Luckily we have a whole rotating team of Edgar's extended kin on hand. It's very hard to tell who are in-laws and who are nephews, aunts, or grandkids in the Winters-Flewellen *Brady Bunch* of a clan. Especially since a few in attendance might be locals Margaret met at the feed store that afternoon. But Margaret's got a good work ethic–detection meter. Everyone is always uniformly pleasant and hardworking if they make it to this remote farm, which proves vital. Each time someone gashes a wrist to an extent requiring bandaging, a new farmer will crawl under the seed drill. So at least we get to be close to the soil this morning.

After 20 grunt-filled minutes, when we actually expose the clogged chutes, we have to ram through, hose down, and completely dry their interiors so they don't clump right back up with our own seeds. Crouched in the soil and grunting, we resemble those clever chimps who poke sticks through logs to extract termites. Observing all this journalistically for a moment, I recall thinking, *Man, all we are doing is trying to sink a couple hundred thousand seeds into soft soil. It's about the oldest thing we do, right after hunting and gathering. Have we perhaps been overthinking it these past few centuries?*

9:36 a.m.: Calibrating the Seed Drill. With chutes unclogged and refastened, now we are ready to calculate just how many seeds we want to get into the ground. We are, in other words, about to answer the universal sixth grader math question: "When will I ever use this in real life?"

By adjusting the drive wheel, you determine the rate and depth at which the seed falls into the soil as the tractor pulls the seed drill across your field. It's circular, so geometry is involved. Your field is graphical and angular, so trigonometry plays in. And it's a seed drill, so advanced knowledge of chaos theory is helpful. The decisions you make now will affect your whole season.

Say you want to plant one hempseed every 30 inches in 30-inch-wide rows at a ½ inch depth. First, get out your scratch paper, logarithmic tables, and metric-to-standard conversion charts. Fifty-six minutes later, attempt to move unmarked, rusted levers somewhere beside the drive wheel to correspond with the result of your calculations. A time-saving tip on the math: I find it's not necessary to go more than six places on either side of your decimal point.

Now you can test your calibration. This only involves further jacking up of your seed drill on mushy, uneven soil (precariously, seems to be the custom, on a primitive jack prototype from the Edsel era) in order to measure the circumference of your drive wheel. That will tell you how far the tractor–seed drill caravan must travel for one seed to drop. You're trying to avoid double planting, under-planting, unevenly spaced planting, or planting at the wrong depth.

Of course, the seed exit hole in your hopper will either be too big or too small for your hempseed. So bring several jumbo rolls of duct tape to your regenerative hemp planting as well as a powerful battery-powered drill. These are the easiest and most reliable modes of drive wheel–hole adjustment. Once the seed drops, some variety of metal disk, rolling pin, or flat rubber flap (like a squeegee) follows to cover the seeds. That allows the seed-to-soil contact that hemp wants for germination.

So you've now calculated how much seed to plant per acre. Or have you? A word about those rusty and unlabeled levers you'll eventually find in a hidden spot on the machine. Don't even bother to read the calibration chart that usually resides under the grain hopper lid. For one thing, "hemp" won't be one of your settings. Also the printed drop-rate-setting numbers correlate to nothing but themselves. There are no marked units. If you set your seeding rate to *11* with that particular rusted lever, as on a *Spinal Tap* amplifier, it means "11 on whatever scale the Schmeiser grain drill assembly guy felt was 11 that day."

Might be 30-inch spacing. Might be 300. I think that's so farmers all over the world will be equally confused. It's comforting to know that whether you're planting in acres or hectares, meters or yards, and measuring your seed in pounds, grams, tons, or metric tonnes, at least the system is fair.

Even the John Deere–capped host of a seed drill–calibration video I watched to see if I was exaggerating some of my experiences could only say, "Sometimes you can get some help calculating by reading the calibration chart you'll see under the grain hopper lid. But [and here he pulls out a calculator], it's best to use this chart as an, um, starting point for your real calibration." He should have added, "Also be careful, the heavy grain hopper lid tends to fall off backwards loudly if you have an idea as crazy as trying to open it." I nearly squashed a Vermont colleague in a hopper lid mishap one year.

If your seed drill is relatively small, let's say 10 feet wide like ours in Oregon, it might have a dozen chutes that can empty that many rows at six-inch row spacing. If you want 30-inch row spacing, leave one hole open, then fully duct-tape up the entry point for the next four chutes in the hopper, then leave another one open. Be careful as you tape these tiny spaces: If you leave any sticky part of the tape exposed, you now have a seed planter that works more like a strip of fly paper.

Now you're ready to address your half-inch desired planting depth. It might make you feel better to adjust the particular rusty lever assigned to that task. But doing so doesn't adjust the seed drill in any noticeable way. You wind up dangerously duck-walking alongside your tractor while manhandling the soil-punching disk part of the device until it seems like it's churning less than three inches of your microbe-rich soil.

In case I've disguised the seed drill takeaway, it's "Plan to spend a lot more time than you'd like calibrating, unclogging, or otherwise tinkering with your planting equipment." It's not so much that seed drills won't work. It's just that they almost never work well at first, sort of defeating the time-saving reason for their invention.

After a dozen plantings, I've seen a seed drill calibrate without hassle only twice. That's an .833 batting average. Babe Ruth only hit .342. So five-sixths of the time, something goes wrong enough for stress phero-mones to intrude on the oxytocin. These do not mix well. There should be warning labels to this effect on seed drills.

One of my two seed drill–assisted plantings that was worth the effort came 3,000 miles to the east. It worked primarily because my Vermont partners' neighbor, Charlie Morse, had waded deep into that scary back part of the barn and unearthed a solid 1940s unit. This seed drill came to us from before what my sweetheart calls the Time of Plastic Crap. The thing was all metal, from an era when more pride was taken even in steel forging.

Once the New England branch of the recluse spider family had been relocated and the very few moving parts lubricated, this seed drill required no maintenance. And, icing on the cake, Charlie oiled the delightfully old-school main spring that anchored the device's axles with hempseed oil. Worked great.

The other key factor in that planting's success is that the seed drill was designed before the era of overengineering. Keeping things simple and functional: That's how people survive in northern Vermont.

When Cary laid out our cultivation plans to Charlie over a microbrew one evening before planting day, Charlie puffed a few contemplative drags on his ever-present cherrywood pipe and said, "I think I have something that'll work. But it's old." That turned out to be an understatement, but if I've learned anything from four years working in Vermont, it's that Charlie's is the culture of understatement.

The unit itself was small (the footprint of a golf cart) and simple: Two direct chutes dropped seed from two sensibly side-mounted grain hoppers next to two posts whose cymbal-shaped tops you could grab to manually adjust planting depth. Took half a second. Just a twist of the wrist. The thing looked like nothing so much as a mobile drum kit. There were no fake rubber tubes to clog: The seed dropped from the hopper to the soil. No calibration was required other than a touch of tape here and there—one small part of the right hopper had rusted through. You could tell from a look that this thing was going to work. When they needed refilling, the grain hoppers were not just easily accessible, but their tops came off only when you wanted them to, with a satisfying, sibilant *clang*.

Charlie, who like most Vermont farmers looks to be somewhere between 40 and 90 years old, loaded the seed drill into Cary's truck bed with a winch, took it to a meadow owned by Cary's father-in-law, and by midafternoon our Vermont acreage was planted and I really was tubing the closest river. That's how these Yankees roll. You don't hear a whiner north of the 32nd parallel.

I'll never forget Charlie, champing on his pipe from the tractor seat while the rest of the crew walked beyond the seed drill, raking dirt over the seed: The device had no squeegee, roller, or disk to fill the hole into which the seed dropped. So humans—including Secretary of Agriculture Tebbetts—played that role. Nice day for a stroll.

To the casual observer, the pace of that planting might have looked a little slower than many modern plantings. But it was the exception to the FLOATER rule. We hardly stopped. The 2018 Vermont planting was my first in a continuing series of lessons that older might be better, when it comes to farming modes.

Why did humans ever stop doing it this way? My guess is persuasive fake-rubber-chute salesman. Or else the fake-rubber people bought out the seed drill people.

10:42 a.m.: One Row Planted. Back in Oregon, a seed finally gets planted when and where we request. It's always a fun moment when you're the guy crawling on his belly behind the seed drill procession, and you can shout, "A seed dropped!" In doing this, you employ roughly the same tone of voice as the starving *Santa Maria*'s lookout did when shouting "Land ho!" from the crow's nest.

But even when the seed drill appears calibrated, here's a tip I learned from Dan Townsend in Washington: Check on the seed drill's performance often. Stop the show, get out of the tractor (or hop off the oxen yoke) and crawl around with your eyes attuned to seed drop rate and planting depth. Do this once per pass, even if things seem to be going smoothly. You'll see why.

Once actual seed starts to drop, morale, which has started to droop with each wasted hour, is restored. The importance of this period in the day is not the 11 percent of the field you'll get planted before the next problem arises, but the resumption of bonhomie.

The successful moments in a group planting provide an ideal way to bond with new partners. This is serious entrepreneurial advice. This is when you see who brings the work gloves, physically and spiritually. Who does and doesn't complain about the sun/humidity/smoke/rain/blood. The humans with whom you've shared this hemp birth are your community. When we say, "Keep the economy local," this is why. People vested in their own backyard can lead the rebuilding of rural societies worldwide. You have these people's backs and they have yours. Not sure how often that is the case with stock trading.

11:28 a.m.: Second Seed Drill Malfunction. The most interesting fact about the seed drill tool is that the reason it will fail once per pass, per FLOATER, is never the same twice. For such a relatively simple machine (at the end of the day, it's just a glorified slide), what's astounding is the diversity of what can go wrong.

Maybe the way you set the incomprehensible calibration levels causes half your 20 acres of seed to drop on the first pass (or, after 10 acres, you notice almost no seed has been planted at all—I've experienced both). Sometimes you don't even have the right size of discontinued hitch ball needed to connect the seed drill to your tractor, necessitating another town run before you even leave the barn. Surprisingly often on a day on which you wake thinking you're planting a crop, you find yourself on hold with outsourced customer service in the Balkans. It's different every time.

One of my more harrowing seed drill malfunctions had all hands taking turns huffing stuck seed out of chutes with a sort of homemade blowgun made from a section of garden hose. That's how inaccessible this often-needed part of the machine was. At another planting two of us at a time had to ride, terrified, atop of the moving seed drill, using brooms to force the seed from the hopper into its chutes.

And no FLOATER problem will ever instruct you how to solve a future problem more quickly. The important thing is that something will malfunction and at intervals throughout the morning you'll find yourself stranded in the field on your back, usually when profoundly hungry for lunch.

Our second total seed drill failure of the 2018 planting occurs when the freed chute gaskets—which we spilled so much blood to un-fuse—now don't want to stay reattached to their chutes. Their size, viscosity, or maybe chemical composition has changed. As a result, one or three of them keeps sliding off over the course of a pass, dropping a half acre's worth of seed in one heartbreaking mountain of fecundity. This necessitates breaking out the toxic goop with which organic agriculture seems to always bring one into the closest contact. Awful epoxies and two-part bonding cements make the chutes once again inaccessible. But at least they stay on for a pass or two.

Unless rain is imminent, this is a good time to break for lunch (you won't, for another hour and a half), while your fingers still work and before they become completely covered with noxious gel.

12:05 p.m.: Customer Service Call. Every hemp planting is as unique as the birth of a child, and every one is an education. For instance, May 28, 2018, is the day I learn that you don't want to be on the wrong end of an Edgar Winters customer service call. He speaks his mind, to be sure, starting in the humble Alabama drawl and gathering steam as he gets worked up. But the call is always more confusing than shaming to the subpar contractor or farm-equipment-rental fellow.

This is because Edgar is, with the passing of Yogi Berra and with all due respect to both Ringo Starr and George W. Bush, the world's reigning king of the unintentionally metaphorical malapropism. Especially when you catch him stressed out in the field.

"I've got this thing all jacked off," I hear him inform our seed drill–rental salesman via Margaret's cell phone. "We've lost two hours of the day, and my brother-in-law's still in the field, dismembering the thing."

Ten seconds of silence ensue on our end of the call. Then Edgar says, "Nah, you're missing the . . . You gotta make sure people return your rigs clean, man. It's only logistical. What? Yeah, logistical. Like Dr. Spock. Now the dang gaskets won't even bond to the . . . to the—"

That call only scratches the surface of both my adventures with seed drills and of what I call, in my texts home to family, Edgarisms-of-the-day. Bigger picture—check your equipment carefully before you use it. Modern hemp pioneer Ryan Loflin, a farmer we met in the pages of *Hemp Bound*, said you can bypass farm-rental shops and get planting gear cheap from the federal Farm Service Agency (FSA) office in some regions. Thanks to the hemp provision in the Farm Bill, now you can even tell the good folks there what crop you want it for. You still might want to make sure the chutes are clean before you load it up.

12:12 p.m.: Running Barn Laps. I am panting hard as the mercury crosses the triple-digit threshold just past noon. My legs feel rubbery with each barn run during this chute reassembly phase. (I am in charge of fetching tools.) I am no longer trying to avoid stepping on fellow farmers when I return to the repair spot. I do try to leap across already-planted rows, but I am not really paying attention to where I land. And, ominously, it is starting to cloud over again. A cloudburst might provide relief from the heat, but the tractor can sink up to its axles.

This is when pain, hunger, and close quarters begin to accentuate your annoyance at the way your nearest farming partner breathes. He sort of hisses out of his nose and through his mustache in the most grating way. And the mustache badly needs trimming. It's difficult to tell where nostril hair ends and external facial fair begins.

1:15 p.m.: Lunch, First Aid, White Lies. Limping inside like a team down 17 points at halftime, you find someone has prepared the necessary 9,000-calorie lunch. As in commercial fishing, it's important to cut the cook in for a full share. An army runs on its stomach and all that. Also as in the fishing trade, tall tales are necessary to maintain morale. The cook (in this case, Margaret's mother, Kathy) asks me brightly, "So how's it going out there?"

A just-arrived volunteer is paying attention to my answer. We have spent most of the morning repairing equipment. We are bleeding on the carpet and dripping sweat into the stew. I answer, "Could hardly be better. They say rain's on the way. Perfect for germination." As yet, we have little to germinate.

2:08 p.m.: Bee Distraction. With bellies full, just when we are all ready to really get planting, someone remembers that the remainder of the seed got left back in the garage. The tractor is switched off, which everyone recognizes is a highly experimental maneuver. This is when we hear the bees.

Bees are the new Save the Whales, what with colony-collapse disorder and the growing movement to rid humanity of dangerous synthetic pesticides once and for all. But even before that, everyone from A. A. Milne to the inventor of the facts of life story recognized that there's nothing that says "life is overall good" like a lot of bees zooming around. When engines are off, the digeridoo-esque white noise captures your awareness as completely as a jet passing overhead. The bee density at planting season isn't yet anything like the cloud it will become by flower time in August. But your groggy, yellow-and-black fellow hemp farmers are waking up, working the raspberries, and checking out who you are.

Both honeybees and native bees adore hemp flowers (especially the male flowers). "It's just shocking how valuable hemp is as a pollen resource for all kinds of bees," Colorado State entomologist Dr. Whitney Cranshaw said in a 2018 interview.[1] I once watched a bee whose corbiculae were so loaded with hemp pollen pom-poms that she had to take a running jump to become airborne. I remember thinking, *Pace yourself, sister!*

Bee bliss session endings tend to be abrupt. In our case this afternoon in Oregon, someone must have found and emptied the seed into the drill hopper. I learn this when the tractor clears its throat and Edgar, who has a bit of a bearded, prancing Bombadil presence in a hemp field, shouts, "Hey, are we planting or are we napping?"

I jump away from my bee bush guiltily when his voice rings out, trying to look busy, but discover he is addressing his grandson, Chris, who is doing his meditation over at his own bee bush. And so we set to work on another pass.

4:21 p.m.: Halfway Through Sixth Pit Stop, Contemplating Hand Planting. It is during this final delay (too many seeds dropping from the pesky third chute) that, as I always do at this point in a mechanized planting day, I begin to think about earlier techniques. I wonder, *Can it be as efficient (or close) for a professional, digital-age hemp farmer to hand-plant and hand-harvest as to use tractors and combines?*

If you're not a master mechanic and blacksmith, the answer might be yes. To see why, consider my all-time favorite seed drill FUBAR. It was in 2016 in Vermont, and the seed drill was wider than the roller that followed it. We wanted them to be equal so the seed that dropped from all the chutes would be sufficiently pressed into contact with the soil. That was when John Williamson popped into his shop and, before my eyes, cut and attached two eight-by-six-foot-wide metal sheaths to extend the roller's range.

Unless tossing on a welding mask and building your own equipment is your idea of a morning meditation, you might consider the calculus of hand planting. Because it's really not that hard to press a hempseed down the necessary half inch with your finger. I've hand-planted small-acreage fields in several states, and here's the only calibration tip you need: A half inch is about the length of your finger up to its first knuckle. (In the "This Little Piggy" sequence, I usually use Wee Wee Wee All the Way Home.)

If you have six colleagues, you can probably plant three or four acres a day, budgeting some time for daydreaming and bee watching. Moderate acreage harvesting, too, can be done by hand, as we'll see.

6:23 p.m.: Elation. The seed somehow gets in the soil. You eat dinner overlooking your planted field. And what's more, you have no regrets. Of all the season's ups and downs, planting day almost always provides the fondest memories.

You sleep deeply and dreamlessly following planting. Maybe because you've been absorbing so much vitamin D. Maybe because, like a dog circling a dozen times before settling down on a couch, your body tells you you've taken care of something elemental.

Then, a few days later, before you expect it, someone spots the first green hook of a sprout. Then 10. Then 1,000. Again flows the oxytocin. These are moments you never forget. You watch early cell division and all the associated miracles, followed soon after by the first leaf bifurcation. Before your own hands have healed, you see that distinctive cannabis hand shape forming on the little guys and ladies.

Congratulations, you're a hemp farmer. You're serving your family, community, and species by doing something that 10 years ago could have got you 10 years to life. Now it's extending your life, and hopefully helping the customers who wind up enjoying the harvest.

CHAPTER SIX

On Weeding, Watering, and Organic Certification

Oregon, 2018

*This is what I love about flowers. Wherever possible,
they just grow; in between the weeds . . . so confident of
their short-lived beauty.*

—Asma Naqi

Even once they've irreversibly launched into planting what
they want to grow, farmers love to second-guess their plans
for the plants that they might not want to grow. In Oregon
in 2018, it was Edgar's uncle Randy who brought up weeding protocol.

Randy was our A-team tractor driver, and at that moment was rev-
ving the vehicle's engine loudly to get it unstuck from a patch of midfield
quicksand. So he asked me, in what you might call his outdoor voice,
if I agreed that the previous week's foot-deep tilling of the interwoven
pasture grasses, prickly lettuce, and clover that had been lining the field
for two decades was the right way to go.

"Yep," I screamed.

That was a specific answer for a specific field. The core weeding question to research and answer before you head into planting day in any
ecosystem is, Should you think of all plant species other than your hemp
as enemies? Or are some of them more "sister crops" (or even "companion

plants") than "weeds"? For regenerative farmers, another way of posing this question is, Do you assiduously yank or chop anything that might "compete" with your hemp following planting, or do you rediscover the benefits of polyculture?

Overall, I've had best results by implementing the latter philosophy. In Vermont in 2018, our monster Samurai plants—some 12 feet tall—were surrounded by milkweed, which helps sustain our root zone biome. In many rows that year, you would find a dozen or two of the foot-and-a-half-tall podded crop oozing milk and silk on you as you cartwheeled by. Sometimes they were packed in tight clusters—virtual forests of milkweed. The hemp plants loved these friends, obviously. Plus we humans benefited from the age-reversing practice of dodging drunken monarch butterflies all season.

Out on Colorado's Western Slope, similarly, the Salt Creek crew encourages tiny clover forests to cluster in the shade of its sinsemilla hemp crop. These nitrogen fixers have the three-tiered role of (1) feeding the soil, (2) attracting bees, and (3) distracting grasshoppers from the hemp plants. In Aaron's words, "We were the only field on the Slope that I know of that didn't have grasshopper issues last season."

Some folks are into permaculture principles, which I encourage. I'm going to sidestep the "till versus no-till" discussion except to say that while I try to treat soil gently, in my experience, true no-till techniques will only work well in healthy soil. If your soil is stressed, your crop is going to want as much help as possible.

My New Mexico home soil being healthy and the acreage small, for instance, I hand-planted on the ranch in 2019, barely moving the crusty soil culture except as necessary to midwife a seed a half inch under the surface, to be babysat by worms. As I examined the horehound's forking maze of omnidirectional roots in that soil earlier today (my fingers are still horehound redolent as I type, mixed with a little globe mallow), I was grateful for its work aiding my family's well-being by aerating the soil. That's free interior decorating for beneficial microbes.

Let's not get too dreamily crunchy about the weed discussion, though. Farmers have to be pragmatic. (That might be the one-sentence summary of this book.) For generations, professional farmers like Dan Townsend in Washington have lived in abject fear of the region's prolific if beautiful yellow-blossomed mullein stalks getting ahead of the money crop. Listening to Dan and his brother talk about mullein is like reading about pre-revolution Russians speaking of wolves. Legends of Were-Mulleins scare Pacific Northwest children around campfires. And they do have amazingly durable, extensive roots.

Closer to the equator in Hawaii, each time I joined Kimo Simpliciano in his moringa fields, I relearned that agriculture in the subtropics can require not just daily, but twice-, sometimes thrice-daily weeding.

"Your target crop grows into trees in the subtropics, but so does everything," Kimo taught me. It's true. Nearly everyone in Hawaii has a fruiting, 11-foot papaya in her yard from a seed she tossed over her shoulder four years ago.

It is borderline terrifying to see how much plant mass can accumulate over the course of an afternoon. One time before his fire, I actually took a morning and afternoon photo in Kimo's field to prove we had weeded the field earlier that same day.

Vigilant weeding is particularly vital when your plants are still *keikis*, which is the Hawaiian word for young kiddos of both animal and plant species. The Ogham language, similarly, uses plant- and tree-based orthography even in its word for *people*—both in this case derived from the same linguistic root. I wasn't aware of either of these historical tidbits at the time my sons were born, but we gave both of them plant middle names.

That's all very nice and on theme, but I still feel for my colleague Melody Heidel, the coordinator on a University of Hawaii project for which I served as an affiliated researcher, who was in the field every day, boxing guinea grass and morning glory.

"With no real cold season at our latitude," Heidel said, "potential competition from weeds can be continuous, so we have to keep them check a bit, especially when they're keikis."

Plant your seed and the race is on. The phenomenon occurs not just in the subtropics, and not just with weeds; by August both our Oregon and Vermont crops were growing an inch and a half per day, almost visibly. This is more daily growth than some adult humans experience emotionally in their whole lives. Once you get into a sort of alpha state from the bee symphony, staring at the plants midseason is like watching a minute hand.

I suggest making hand-weeding at least part of your strategy, even up to 10 or 20 acres. For one thing, once your hemp is established, it proves a dang solid weed suppressor itself. Meaning, the work is most intense early in the season, when you're fresh and the visibility is better. You'll be in the field anyway, per Bill Althouse's advice. You might as well do some isometric bending. Weeding is relaxing—another excuse to be outside a little longer.

In addition to weeding, farmers also like to overthink and tweak their watering strategies. This is perhaps to be expected. A plant, like a human, is 60 to 80 percent water. We sense the importance of having a watering plan, but we tend to be helicopter parents. Once perhaps three-quarters of a field is planted, the farmer mind starts leaping ahead to everything that has to be done once it's 100 percent planted. Of course, how (or if) you want to water should be dialed in well before planting day. But that doesn't mean you'll ever stop debating how to keep those roots moist (but not too moist) after planting.

By this point with various projects, I've used just about every watering mode, from sprinklers supplied by a well in Oregon to acres-long pivots raining Columbia River water on Colville hemp via huge, rickety old pumps. In New Mexico I'm running a solar-powered, gravity-fed drip line with supplemental hand watering as necessary in the plants' early vegetative (preflowering) stage. And I've dry-cropped every year in Vermont, but that's like saying Alaskan ski resorts don't use artificial snow: It's wet enough from the sky.

Planting, unlike harvest, can take place in a light rain. It won't sink your tractor. But it feels like supergood karma when it rains on the day *after* planting, as happened two years running in the Green Mountain State. Knowing this, you tend to finish up planting days by doing the same rain dance farmers have done for 8,000 years. Be sure to conjure steady, moderate rain to facilitate germination, not millennial floods.

Hemp likes moisture at germination, but doesn't need a bath. Just dampness at that magic moment of seed-soil contact. After that, it doesn't want its feet too wet. Especially once flowering begins. But even at germination, more folks, myself included, tend to overwater than underwater. In short, drop the seed in the ground, moisten it for a day or two, and the hemp will know what to do. If it's good seed and decent soil, it'll come up.

Flowering is stimulated in a hemp plant after the summer solstice, when the days begin getting shorter. Once your keikis are up and established, the best course for watering throughout the vegetative stage (or *veg* if you want to sound hip) is once again to follow Bill Althouse's counsel and carefully watch the crop. If the leaves are drooping a bit, the plant is thirsty. In some climates this will be almost daily. In some, much less frequently. It can depend on the time of year too. In Colorado, fields need much more water in July than in June. Just one state south, in New Mexico, it's the opposite, so long as our monsoon rains come on time.

Back in May 2018, at planting time, Edgar had a few last-second thoughts about where Chris had positioned the sprinkler heads that were mounted to fence posts at various points in the field. So, with minimal whining, we repositioned a half dozen of these just when the seed drill and tractor were, for the moment, working. On average, we wound up watering that field about every third day.

That Oregon hemp canopy kept the crop shockingly moist even during the peak of the 2018 wildfire, when the rest of the region was panting. I learned this around midseason, when Edgar grabbed my arm one afternoon while I was trying to determine the gender of a plant and said, "I've got something for you to feast your fingers on."

What he was showing me was how wet our soil was, compared to the ground exposed to the flame-parched world outside the range of our crop. In fact the very air was easily 10 degrees cooler at ground level inside the hemp field than outside it.

In an aerial photo taken this day, I can see how dry everything around the field was. In fact I can actually see the wildfire smoke in the background. Our soil stayed moist the whole season, long after the leaf canopy prevented the sprinklers from getting much water to the base of the plants. The hemp was protecting an entire ecosystem, including us. Right through harvest, Edgar, Chris, and I used to crouch down in the lush field just to get a break from the smoky haze.

Outside of water and weeding protocol adjustments, a third important element to keep in mind before planting day is organic certification paperwork. Usually a large part of this is documenting your field's planting history and soil amendments. For that first organic certification in Vermont, our field had no history of amendments or planting, which made things a lot easier. Generally speaking, organic rules mandate that prior to certification, three years must pass since any nonorganic pesticide or herbicide applications.

Each state has its own mode for organic certification, and there are also third-party certifiers who will travel to you if you want to go that route. Start with your state department of agriculture website to find out how to get rolling. In some states the paperwork can be a pain (it was relatively easy in Vermont), but it's worth it.

When it comes to applying anything to soil or plants, without question, cultivate your hemp in an organic fashion from your first crop, regardless of whether you go for USDA certification. I recommend you do: On top of the value the certification adds to your harvest, organic products are better for all critters, microscopic and macroscopic.

Without delving too deeply into the legitimate debate about the viability of any federal standard designed by lawyers to fit 300 million consumers, at least organic certification means you are avoiding some of the more disastrous chemicals designed for monoculture over the past century.[1] As Nutiva's John Roulac reminded me recently, industrial agricultural runoff is the number one polluter of oceans.

———————

Something that becomes clearer to me every season and might be helpful to keep in mind as you approach your final weeding, watering, and organic soil prep strategy is this: You'll learn your own field. It may take more than one season. But you'll probably still get a decent harvest the first time out. Think soil health, work with a sense of gratitude that you're not in an office, and the plants will follow.

As a midwife, I'm the kind of guy who likes to thank the plants, verbally, for all the oxygen I'm breathing, all the carbon they're sequestering, all the good health I'm enjoying while tending to them. But it's not much of a stretch to say that as an outdoor hemp farmer, you deserve some thanks too.

Edgar, as usual, said it best a few days after the 2018 Oregon planting, when the first keikis had appeared, our winter-atrophied muscles had turned their groaning down a notch, and we had retired to the pond adjacent to our field.

"Well," he observed, feeding a handful of hempseed to the koi. "That's another milestone in our caps."

Whether you account for your plants as individuals or by the hundreds of acres, every hemp crop is sacred. It builds you while you build soil. Heading back to the Funky Butte Ranch following four 2018 plantings that spanned 70 degrees of longitude, I felt physically strong and mentally healthy. Which is to say, ready for the next series of problems. The midseason ones.

CHAPTER SEVEN

The Midseason Panic Attack: Rookie Focus

Colville Tribal Land, near Omak, Washington, 2017

I've found, as a general rule of life, that the things you think are going to be the scaliest nearly always turn out not so bad after all.

—P. G. WODEHOUSE, *The Inimitable Jeeves*

I t's usually around 60 days into the season that a farmer jerks awake in the night, panting with a premonition about something biblical going wrong in the field. Might be hail, deer invasion, or state inspectors insisting on testing cola flowers.

Last summer a colleague shot me the most adorable picture featuring a nest full of just-hatched thrush of some kind. They were asking for breakfast from the middle of the hemp field. The adorableness wasn't the reason for the note. At least not the only reason. She also wondered if the family's superfood dining habits might prove a significant detriment to the harvest.

You're going to have a midseason panic attack surrounding your first hemp crop. It's just probably not going to be for the right reasons. That's because the debut season hemp farmer usually freaks out only about the field, rather than also thinking about the great unknown that lies beyond harvest.

For instance, one morning late in the summer of 2017, I woke to a series of horticultural text images from Eastern Washington, accompanied by a request to report to the Colville field to investigate. My heart was heavy as I sipped my hemp, turmeric, and blueberry shake that morning. For one thing, few demands of modern life are more grating than trying to sufficiently enlarge a phone-transmitted image to the size one would need to begin to make an actual diagnosis about anything botanical. Was that an aphid or someone's elbow?

What became clear was that in at least portions of the tribe's 60-acre field, the outer perimeter of the leaves on the just-starting-to-flower field were turning yellow in a sort of jagged pattern. Suddenly I was viscerally experiencing the spot-on nature of Bill's advice about the ceaseless vigilance required when your living depends on Mother Nature. I was not dirty at this moment. But I wanted to be.

If a crop is showing evidence of distress by August, it's not necessarily too late to do something about it. And you should address the issue, even though it's almost definitely the result of preseason soil realities or genetics. The alarm itself is what matters, not the cause. Whether or not the steps you, the helicopter parent, take to deal with the field issue are the catalyst, that field issue tends to get resolved. It feels like a code red at the time, but you'll live to fight another day.

Partly that's because you can often treat the symptoms, say, a crop that is late to transition from veg to flower. Thirst is generally the easiest plant message to understand and mitigate. Pests, weeds, or unauthorized critter visits are next easiest. Nutrient surplus or deficiency is harder. And genetics, being already in the ground, are impossible to mitigate, until next season. In Colville that season it seemed to be a nutrient issue plus the crop's underlying genetics.

But the real reason to take a field issue in stride is that plants are smart. That might sound like a compliment, to some ears even a woo-woo declaration. But when you give it some thought, it's fundamentally condescending. Plants know they are clever. Ferns, to name one family, have been humming along without fail for 360 million years, 120 million years longer than we Johnny-come-lately mammals have been around. They don't need our affirmation. Plus, we'd be pretty offended if the first line of a chapter in some plant author's book about us read, "Many of the more hairless primates are not the absolute imbeciles their post-shamanistic leadership style would indicate."

Still, recognizing the scope of plant intelligence reflects a fairly new mind-set for me. It provides immediately valuable information: Plants, like most intelligent life-forms, are good communicators. They deal with a problem if they can, and if necessary, they tell us very clearly how things are going in their lives. I wouldn't say they gossip, but if your plants are not happy, you'll know. They are without question chatting. With each other, with the microbe community we've been architecting, and with us.

That plants are talking to us is not new information for the world's farmers, historically. Certainly not to Bill Althouse. He's part of a tradition that in the United States goes back to colonial times. George Washington knew to tell his head gardener of the hemp that he grew on all five of his estates, "Let the ground be well prepared."[1] General Washington understood that what comes out of the ground is contingent upon what steps you take before planting.

It was only recently that this commonsense awareness was nearly forgotten. Plants and animals split. About the time that the "12 items or fewer" lane came online. But like separated twins living similar lives, we've always known deep down we were related.

Like most kids, even growing up in 1980s Long Island when Madonna was like a virgin, I wondered if trees "feel." I mean, I recognized as I swung a hatchet that it is a powerful act. I grew up at a time when you could say, "talking to plants helps" and not be considered insane. You could find the occasional middlebrow journalism article to back you up.

But when it comes to totemic relationships, I'd always been more of an animal guy. I'm the fellow on the rain forest canoe trip who woke early and locked eyes with the jaguar mother and kitten having a sunrise drink. In everyday ranch life, I'm the goat whisperer other members of my family call down to the corral if Bjork is acting rambunctious on the milk stand. It is my sweetheart and kiddos who have the green thumbs in the family.

It's taken two decades of journalism, first about local living and then about the cannabis/hemp plant, to provide the immersive education that's allowed a much muddier me to truly understand the sophisticated intelligence of, for instance, your basic bean sprout. Now I see clear as day that plants, as they go about their day, do the equivalent of getting the chores done, feeding the family, and squeezing in some exercise. In short, plants do what we do.

Ultimately the only substantive difference between animals and plants is one of pace. Plants divide their year seasonally, rather than weekly. This awareness has deeply affected nearly every phase of my life. You should hear my Chauncey Gardiner–like requests for journalistic deadline extensions these days. I hear myself saying thing like, "The piece will ripen in due course."

Likewise my entrepreneurial philosophy is now grounded on the pace taught by plants—I model business plans on creatures that grow for 10 months and allow you to make a living while working on climate stabilization. Advance soil building, high-quality hemp-processing modes that require more time, caution about the chemical composition of the people with whom I consider collaboration: I've learned from green leafy cannabis plants how to handle all of these. And it seems to be working out.

In the ground, of course, is where plants teach the most directly. Three months before planting hemp here on the Funky Butte Ranch, I began my overwinter soil preparation by planting a nitrogen-fixing clover-vetch combination. A crop that has, on the surface, nothing to do with hemp, is going to one day soon play a role in the hempseed shake that I and my family enjoy every morning.

However it works in the real work marketplace, I find that patience as a business strategy suits my nature. Throughout the hemp-cultivation cycle, I keep relearning the same lesson: that best practices might be about leaving a few steps out. That fifth isn't the only gear for a farmer, entrepreneur, or species aiming for survival.

But that awareness not to freak out the moment you have an issue in the field has been hard won; it's not how I handled it in that 2017 large-acreage field. In truth, we didn't primarily have a field problem in Colville that year. We had a harvest-transport and drying problem. But we wouldn't be aware of this for several more months.

You never expect a negative diagnosis. When those yellow-leaf text images came microscopically over the ether, I, like most relative newbies, had tunnel vision. After spending the day overreacting and booking flights, I crashed but quickly woke from nightmares about regiments of diseased leaves—suffering from much more severe disfigurement than the real leaves—storming in formation after Jackie and me.

I cared too much about that Colville crop. I really wanted it to succeed. Even the Colville Tribe's chairman at the time, Dr. Michael Marchand, had told me that the tribe was deeply committed to hemp. "We always thought, maybe this is the crop for us someday," he said. "We want to diversify our economies. Hemp looks like the place we want to be. I'm happy to actually see a plant growing."

There was, in other words, a lot riding on this crop. As lead consultant to the project, it was on me to figure this out. In waking life, all I wanted to do was change those leaves from yellow back to green. I wanted a magic wand. A magic soil supplement. As for jagged yellow leaf perimeters, I had never seen this variegated pattern before. I found myself taking it personally.

Hmmm, I thought. *Could be overwatering. Then again, could be underwatering. Could be too much nitrogen. But dang, given the struggling soil when we planted the crop, could be too little nitrogen.*

That first morning with a midseason problem, I perused the Emerald Triangle ganja blogs that I find provide the most comprehensive photographic analysis of cannabis issues. To this day, I love sending former black market cannabis-cultivation URLs to clients as part of my official advice. Best gardeners of Michael Pollan's generation, indeed.

Hmm. This just added to my variables. I scanned every yellow cannabis leaf since prohibition began. Looked like it probably wasn't a potash issue. The third-generation greenbud experts said leaf discoloration in that case is often associated with spotting rather than striping. That's when it occurred to me to FaceTime with someone who had 60 years of cultivation under his belt. It's good to have the world's most experienced hemp farmer as your mentor.

"Sounds like a real quackmire," Edgar said when I had forwarded him the tiny photos of apparently struggling plants. "In year one, when you're building stressed-out soil, this kind of thing can happen."

When I asked him if it was more likely a nutrient issue or a genetics issue, Edgar replied, "Could be either. It's a half dozen of one, fifty of the other. These problems can cohabitate."

Michael LaBelle, when I discussed the quackmire with him two years after the fact, concurred. "Pictures might be worth a thousand words," he said. "But on-site time is worth a thousand pictures." In other words, diagnosing a field issue is not so easy until you see it in person. Especially a large-field issue.

I was learning that fast. A key question to ask with any plant health problem is whether it's localized or pervasive. In Colville in 2017, it was somewhere in between. At the time, Edgar and I agreed a sulfur deficiency was the likely culprit, based on preseason soil testing and the outer-edge yellowing pattern.

Jackie was on board for a sulfur application. But adding OMRI-compliant nutrients to 60 acres in the middle of the growing season is no small task. That was the week I went from gentleman hand-weeding kale gardener to pricing aerial sulfur applications and back. We landed somewhere in the middle. Jackie decided to add the nutrients to the pivot watering apparatus in solution.

But really, the core problem in the Colville field that year wasn't sulfur any more than the problems in the Middle East are about last week's headlines. By the time I sashayed to the field with my family and dogs in the ol' camper, after driving through the—ho hum—Armageddon of wildfires the whole way through Colorado, Utah, and Idaho, the crop already looked much better.[2]

Wading each morning into a square mile of Kelly green hemp, scaring the bejeezus out of the sun-darkening flock of blackbirds and neighboring rancher's goats snacking on the crop's just-forming hempseed, my sons and I had to get deep into the weakest portions of the shoulder-high field before I saw them: fast-maturing leaves yellowing in a weird pattern around their perimeters.

I thought, naturally enough, that the sulfur application had helped. But when I spoke to her about it more recently, Jackie, who knows the field best, said she doesn't think midseason liquid sulfur application was the way to go. Indeed she isn't convinced sulfur was the main issue that year.

"Obviously the ideal situation is to address your nutrient shortages months before planting," she said. "But if you have to do a midseason sulfur application, granular rather than liquid is less likely to burn the plants. And whether or not sulfur was our main issue, you can't expect to monkey with your nutrients too much midseason."

Indeed Dan had already noticed that his prior season cattle run sported by far the best swath of the field. That's where we took all our social media shots that season. So something was happening with nitrogen too. Suddenly I was thinking it was a good thing the state of Washington's crappy genetics laws didn't allow the tribe to save seeds in 2017.

Which really gets to the heart of the rookie midseason focus: Thankfully the crop did improve, and did get harvested. But even if it hadn't, Jackie could only do her best to mitigate what was at core an off-field issue: bad genetics policy. Yellowing leaves had their origins in disastrous hemp legislation passed two years earlier in Olympia, Washington: no domestic cultivars allowed. Late-arriving permits delayed planting. Only four projects even applied to the program that year.

The situation in Colville in 2017 is the reason the term *perfect storm* was adapted to nonmeteorological phenomena. In fact, weather was about the only thing that went right, other than perseverance. The tribal project was forced to plant non-acclimated seeds from tired genetics sporting a germination rate lower than I would ever sell. Furthermore, a fiesta of bureaucratic nonsense forced the project to plant in July instead of Jackie's target date of May 20. Robust hemp genetics planted in early June could've handled the challenging soil situation. The law and regs were so unfriendly that it was something close to a miracle that, after months of paperwork delay and weeks of confusion at customs, the sacks of European seeds even arrived in Seattle. All for genetics that the project never should have been forced to import in the first place.

With the ton of seed actually in the United States, it looked like planting might finally be upon us. So I flew in to Spokane right around the July 4 holiday. For 3 days Jackie begged for the release of her project's seed from customs. I don't think I'll ever forget the afternoon of July 8, bumping in Jackie's truck, with its tribal license plates, toward an absurdly out-of-the-way location in Airway Heights two and a half hours from the field, as specified by state regulators. This is where the seeds would have to be stored—if they ever made it out of customs—in DEA-compliant lockers until we picked them up.

Even as we drove, the UPS folks were saying that there were no fleet vehicles in the entire metropolis of Seattle available to get the seed to Airway Heights for the long-delayed planting. As she argued with customer service on the truck's speakerphone in the remote town of Omak, population 4,787 and home of the annual Suicide Race indigenous horse competition, not one, not two, but three UPS trucks ambled past us. Brown, shiny, and in formation.

As a final insult, when we picked up the seed many hours later (technically the next day) and sped off back to the field, Jackie got one last call from a customs official.

"Don't plant yet," the voice said. "We have some more questions."

In the passenger seat, I made the downward "in the hole" arm gesture from *Caddyshack*.

"Too late," Jackie reported. "The seeds are already in the ground. Sorry."

That crop only got planted because a lot of people went the extra mile, including one Victor Shaul of the Washington Department of Agriculture. He, in fact, went the extra 200 miles, on a Saturday, to bring the "second key" we needed to open the DEA-approved seed-storage locker in Airway Heights, from his office in Yakima.

This protocol ensured that, as of 2017, hempseed was as difficult to obtain in Washington as it would be for a president to launch a nuclear strike. It was far easier to grow psychoactive cannabis than hemp in the Evergreen State that year. By contrast, in 2019 I just shipped my Samurai seed home to myself in New Mexico.

If we hadn't gotten that seed out of that Airway Heights storage locker on July 8, it's doubtful the tribe would have planted that year. The days were already getting shorter, for crying out loud. It was nearly two months later than the ideal planting time. The plants were going to want to start flowering before they even matured. Short of Superman reversing the earth's spin, there was nothing we could do about that.

Plus it really was bottom-of-the-barrel seed. Long story, but thanks to the fierce loyalty of my sister-from-another-mother Hana Gabrielová in the Czech Republic, who essentially broke a Canadian seed blockade of Washington that season and sent us what she had left so late in the season, we were allowed to be stuck with offshore genetics.

The French breeders of the cultivar Colville grew that year, a variety called Ferimon 12, didn't really care if we replanted. It was very difficult to find them, actually, and my high school French isn't very good. But the State of Washington's hemp rules that year wouldn't let us save seed anyway. Nor grow for flower. You see why I'm so passionate about the genetic level playing field: I've lived the horrors of trying to plant without it.

But miracles happen, the crop got planted, and it came up. And everyone was psyched. Even the Colville Tribal Council came to bless the crop—an hour drive from tribal headquarters, and not an easy drive. The elders came in a van.

It was an impressive ceremony. There was drumming and chanting. Then Chairman Marchand told me something that proved prophetic:

"This is nice, being here today," he said, gesturing over the vast field. "But once we're gone and the plants are left in the ground, then you're under pressure. It's like any farming: It kinda [chuckle] takes control of you."

Darn tootin'. We don't think about it too much, but a lot has to go perfectly right in life for us to reach our current age. So when we got some leaf yellowing a month later, we all freaked out a bit. But it could have been worse. We treated the symptoms and, I'm only being factual here, kept praying over the crop long after the council ceremony had concluded. On reflection, I think that might have helped as much as any soil applications.

Given the comedy of errors the project faced, I look back on the 2017 Colville crop as a not-at-all-bad debut effort. The tribe wound up with a multi-ton harvest that year. And it would have been even more if we had known what to focus on even as we discussed leaf coloring. But, hey, they say pain is the best teacher. We sure learned. And now you get to benefit from that trial by fire.

CHAPTER EIGHT

The Midseason Panic Attack: Veteran Focus

The Haggunenons of Vicissitus Three have the most impatient chromosomes in the Galaxy. Whereas most races are content to evolve slowly and carefully over thousands of generations, discarding a prehensile toe here, nervously hazarding another nostril there, the Haggunenons . . . will frequently evolve several times over lunch.

—Douglas Adams,
The Restaurant at the End of the Universe

Now that we've covered what not to obsess over when the midseason panic is triggered, let's examine where, if we're prepared, we might shift our attention when we wake up panting. Psychologists call this phenomenon *attribute substitution*. It's based on the principle that you can't take the worry gene out of the primate but you can rewire what it is the primate worries about. When it comes down to it, there are six main ways to work off your August stress.

1. Make sure your crop's drying, cleaning, and storage processes are dialed in. The scariest part of harvest, and arguably the most time-sensitive, is not the harvest itself. It's the 2 hours following harvest.

This might be the most overlooked component of the hemp season I've noticed in farmers for whom hemp is their first commercial crop.

In 2017, Jackie booked a drying and cleaning facility for the crop well in advance of harvest. Because the place was (like everything that year) 2 hours from the field, Jackie procured the longest flatbed truck I had ever seen. Dan dumped the hemp from the combine into wooden crates that were immediately loaded on to the flatbed. Still, transporting the full harvest required two trips because, well, there was a lot of it. That wound up costing the project a couple hundred pounds of seed.

Here's the must-know piece of intelligence: The moment you harvest your hempseed, the clock is running to get it down to 8 percent moisture. That's because the instant your seed leaves the field and is deposited in a bag or silo, it starts to compost.

"The reason that happens is that hemp is harvested greener than many grain crops," Roger Gussiaas said. "Getting it dry is your first priority."

Your initial harvest is a mass of green chaff: It looks like lawn shavings. Sometimes you can barely tell there are seeds in there. The crop can come out of the field at 20 percent moisture or higher.

That's a lot of water. Try sticking your arm in a batch of hemp 45 minutes after harvest: It will be burning hot. I'm talking about nearly too hot to stand. You'll be steam-scalded to your clavicle. Your harvest is cooking. Sautéing. Immediately.

Now it's a sprint: You have an hour or less to get going on the drying process. You might consider getting a siren to mount atop your farm truck if you have to move your harvest to your drying location. It's firefighter-mentality time. And it's why the farmer's first midseason task is to dial in the whole drying and cleaning process well before harvesttime.

For your seed harvest, there are lots of ways to do this. In 2018 in Vermont, we used a nearby organic popcorn farm's drying facility. Solar-powered too. Talk about on brand. But not everyone has access to popcorn facilities. Clearly not most modern movie theaters. What most everyone does have access to are large fans. And these, alongside some sort of storage bin (like a small grain silo) and a decent-quality moisture tester, are what you'll need. Adding fun for the nascent hemp farmer is that for now all but one brand of moisture tester—the Dickey-John

Grain Analysis Computer—lack a calibration level for hemp. And the Dickey-John is expensive, around $4,200.

So if you and your harvest team get hold of the much-cheaper handheld field models (which start at around $250), you get to poll one another about similarly sized seeds and then average out the results of several settings: barley, corn, and amaranth, or whatever. I usually find the readings are pretty close to one another. Close enough.

Here again I was lucky in that I learned from the uber-experienced John Williamson how forced-air drying in a silo works.[1] It's not complicated. Basically, make sure you have that silo on-site or very, very close to your field. Depending on the size of your crop, it doesn't have to be huge—just able to hold about 1,000 pounds of seed per acre of field, plus associated chaff. The silo has an open (grated) floor. Below the floor, place a huge fan. It only has to be heated air if your harvest climate is cold and humid.

Once you (or your combine or your oxen) deposit the wet harvest mass in the silo,[2] snap on the fan and take a deep breath: You have started the drying process. Then spend the next 24 to 36 hours turning over your green mass with your shovel, unless your silo is fancy enough to have an automatic churning mechanism. Do this hourly. In Vermont in 2016, I loved hopping fully into the silo, breathing terpenes, and rotating the deep hemp carpet.[3]

The goal is to get the seed dry enough before it overheats and rots into a sterile mash. It's simply amazing how fast this happens. Within 2 hours of harvest, the tribal project lost that entire crateful of seed in 2017. So if you have to go off-site to your own or a commercial drying facility, you'd better get your entire harvest there, fast. That experience is why I believe on-site drying facilities are a worthwhile investment.

There's one more step once your seed is dry to 8 percent moisture: Get it into the seed cleaner. A seed cleaner is a simple multilevel agitating machine that resembles an air hockey table. It shakes like a Polaroid picture, shrugging off the chaff through progressively smaller screens; after you run your harvest through it once or thrice, you end up with squeaky-clean seed.

Now your seed is ready to be replanted, sold whole, or turned into pricey hemp hearts, omega-rich hemp oil, and my goats' favorite: hemp-protein powder. More immediately, your seed is now ready for storage in the temperature-, moisture- and rodent-controlled facility you've been preparing instead of worrying about sulfur.

Today, unless you splurge for jute bags and until someone markets biofiber grain bags, there aren't many great options for storage sacking. Some folks like vacuum-sealing their product (especially flower) in nitrogen within Mylar bags. That's not my personal storage choice. At every stage of the farming cycle, I like to ask, "How would the shaman have done this?" And Mylar just rarely comes up. But it's hard to avoid the woven plastic fiber of one-ton grain bags. They do the trick but, ya know, *yuck*. More plastic junk around the farm. I can't wait to wean from these. Still, once the seeds are dried, cleaned, and stored, your harvest is safe and you can eat some waffles.

For her part, Jackie had a great perspective about learning this and all the project's early lessons the hard way.

"We've had two extra seasons to learn how to grow and harvest hemp," she told me of the tribe's head start over nearly anyone else in Washington. "Sometimes I get caught up in the fact that I want it to be successful and do these awesome things for the earth. Then I remember, we're learning." And sure enough, with the passage of the 2018 Farm Bill, the tribe is now in control of its hemp destiny.

2. Conduct your own initial THC tests, prior to your official state program test. August is also the time when a young 'un's thoughts turn to confusing questions about THC testing. Until THC worry goes away for farmers, expect to be awakened with some form of the call I received from a colleague midseason in 2018 that said, "Does 0.33 percent mean 0.3 percent? The state should round down, right?"

First of all, if you think you could be on the cusp of a hot test result, congratulate yourself for being a good farmer. At the same time, particularly if your crop looks and smells really good, you're wise to start doing your own testing well before harvest. Just so you have a handle on how your cannabinoid profile is developing.

Test in the morning. Test in the evening. Test flowers. Test leaves. Test your big plants. Test your small plants. Lobby for leaf testing. Go to churches and explain why not to be afraid of microlevels of THC.

Some might be guessing that all this testing can get expensive. Indeed it can. If you have a good relationship with a testing lab (as we do in the

Vermont project with a regional outfit called MCR Labs), you can keep each test under $100.

If you have a lot invested in your crop (and who doesn't?), it might be time to take the plunge and invest in one of the newfangled handheld testing machines. These start at about $700. They aren't as accurate as lab-level machinery, but they give you an idea of your cannabinoid levels, and once you own the gear, you can conduct as many tests as you like. Plus you can check your test against a professional lab's (or an official state test's) result.

And for those of us who really love having something about which to panic, here's a great one: What if even your midseason tests start creeping close to that (soon extinct) 0.3 percent level? Now you've got some tough decisions to make. Do you hobble your plants so that they don't risk destruction due to inane laws? Do you (gasp) consider harvesting early and subjecting your customers to an inferior product?

There isn't an easy answer to this hopefully retreating dilemma. But as I learned in 2018, everyone has to be prepared to face this quackmire: The same cultivar can test five times higher in THC in one location versus another. That's what happened with Samurai that year. Only good policy saved our Vermont crop. So test early and often.

3. Gather pollen from your favorite male plants before they die back. While we work to fast-track THC irrelevance in the field, Edgar and I are also working on a breeding program to tame the Samurai a bit. You know, to make hemp cultivation less of an annual edge-of-the-seat adventure. Also we're trying to up its CBC and CBG levels. One way to do this genetics work is by channeling Gregor Mendel, the famous 19th-century monk whose experiments in dominant and recessive genes every second grader re-creates with beans-in-a-cup.

As if you need an excuse to skip around your field dodging butterflies and taking another terpene bath, unicorn hunting is always a fun time of the season. If you're in the dioecious camp, the time to capture male genetics is when your male plants are ripe and ready to pollinate.

By August your males will look completely different from the females. The vertically stacked pollen sacs are the giveaway. Also the bees gobbling them. It's not hard to tell when a male flower is mature—brush against it with the force of a butterfly kiss, and pollen will leap off in a dusty little cloud.

Unicorn hunting involves looking for a specific few plants you like, either because of the way they grow (say, their branching structure, their height, or their lack of height), because of their rapidity of germination, rate of maturation, or because of the color of their fiber. Maybe you've just had a good feeling about a particular plant all season; Edgar's grandson Chris named his favorite plant Henrietta in 2018 and lavished extra love and water on her that season.[4]

To capture hemp pollen, place a small paper or cloth sack over of the male cola flower overnight, bind it at its opening (I use hemp twine, needless to say, but a rubber band can work), and collect the pollen in the morning. Label it (with date, cultivar, location, and what you like about it), and refrigerate it.

If you're growing strictly for a flower harvest (and thus are bringing only females to maturity), by August you'll start "sexing out" your plants several weeks earlier than full male maturity. The reason for this is that you want all your males out of the field or greenhouse before they pollinate.

This is beyond safe sex for plants. Let's call it what it is. You're a dang hemp Puritan when you cultivate this way. But it is what most folks do at this moment in the industry's evolution (assuming they aren't growing clones or "feminized" genetics). That's because it's what the market currently demands if you're trying only to maximize your CBD percentage.

I've become competent at this process, studying the (usually) sixth "node," or branching, from the main stem (counting from the base of the plant) about four to six weeks after planting. You're checking if the first flower nubbin is taking a distinctively male or a female form. Gender is already discernible even in the very early flowering stage, with the female shape more of a branching pair of "bracts" with small hairy stigmata emerging from each, and the male pre-flowers looking like small round sacs.[5] Both are tiny at first, the size of BBs or smaller. This is why I like to examine the young flowers with my pocket scope—plus it looks very professional to whip out a scope in the field. But if I'm decently adept at this process, Margaret Flewellen is a savant. She can sense a plant's gender at a glance, the way a professional poultry sexer can look at an hour-old chick and tell whether it's a pullet or a future rooster.

And as in a chicken yard, where farmers of a spiritual ilk believe that a rooster in the yard makes for happier and more productive hens who lay healthier eggs, you needn't remove your young males right after identifying them. I notice that even in his sinsemilla gardens Edgar shares this belief in a goal of hormonal balance in life. He'll keep some males in the field until near maturity. You have plenty of time—several weeks—to remove the males after ID'ing them.

Now back to our dioecious crops. The reason to "bag" the cola flowers of your favorite unicorns at pollination time is that male plants generally die back after pollination. So this midseason moment is your window for saving genetics with a Y chromosome. Since they depart on their own, I don't usually pull the males after pollination, though Edgar likes to.

Then, in the greenhouse or in the field, you take the step that always reminds me what an artist the great hemp professor Edgar Winters is, and how important it is to own your own genetics.[6] It still blows my mind that this soon-to-be-multibillion-dollar industry hinges on delicately applying individual pollen grains to female hemp flowers with the kind of brush you'd use to detail a model plane. Smaller than you'd use for brushing a toddler's teeth. It's always a surreal experience to do this under Edgar's and Margaret's guidance.

"See how the females here are ready, too?" Edgar will say, looking over my shoulder. "You only have a couple of days' window to get 'em pregnant. The two genders have to be on the same tune."

At the same time we have a quarter of a million plants maturing an inch and a half per day outside, we become this nursery, working to advance our genetics to a place that works for our products. And there's not a test tube or toxic substance involved.

Flower by individual flower, innovation starts with that unicorn that gives you the properties you're seeking. And so an ancient industry is reborn in the digital age, care of an Austrian monk. Spending some time at midseason identifying your favorite plants of both genders is as vital to any agriculture-based endeavor from here on out as water and sunlight and, for that matter, sufficient initial capital.

4. Find a home for the tons of gorgeous product you're about to find piling up in your barn. Say you're taking Dolly's and Wendell's advice and developing a value-added product: If, by August, you're not already out meeting with managers of food co-ops, dispensaries, farmers markets, and supermarkets, well, it's not going to happen magically after harvest.

Whether your product finds a market will almost certainly be directly linked to how devotedly you get out there before harvest to let people know how passionately you believe in your plants, team, and product. The Fat Pig Society, Salt Creek, nearly everyone I know hauled tail for several years before they saw positive revenue flow.

Finding retail outlets is just the start. Have you built a website and social media presence? Figured out how much shipping will cost? Researched the price point at which comparable products are being offered? Learned labeling rules? Splurged for a booth at one or two trade shows?

5. Make Friends with Officialdom, Form Your Company, and Set Up Banking. If your product is going to be eaten by humans, you're wise, by the time of your equivalent of the Sulfur Distraction Incident, to be well into discussions with the nice public servants at the health and food safety departments in your state. Even your processing facility will have to meet certain (and sometimes quirky) design standards for an edible product. And, of course, if you're (wisely) going organic, your paperwork should be done by now as you will be nigh approaching your field visit from your certifiers.

By midseason, as well, you should be talking to lawyers, having your company created, your relationship with partners solidified. Are you a co-op? An LLC? A B corp? How's your mission statement looking? Have you sent a press release to local media about the local jobs you're creating while sequestering carbon with a healthy product?

By 2020 you should be able to obtain crop insurance if you want it, and to bank like any other business. I personally have never had a problem starting a hemp-related bank account, but many of my colleagues have. So it's terrific that this quackmire is going away. But maybe you'll decide to explore local credit unions or cryptocurrency options.

6. Finalize Product Design. Also by midseason, you'll want to be wrapping up work with graphic designers so that your logo and labeling are set up. The look of your product is your perpetual commercial. It's very important.

I find graphic designers vary greatly in the timeliness of their work. Some will bring your amazing logo ideas to life the next day, some the next year. And at about this point in the summer you should be ordering your recyclable bottles and compostable packaging. Leave a space on these labels for batch numbers: You've got to be able to trace the source of every bottle.

Even if you're planning on wholesaling some or all of your flower harvest, now is the time to establish relationships with possible toll processing partners. Some of these outfits are terrific, like Sub-Zero Extracts, and some, as we touched on, are rip-off artists who offer low prices or want half your harvest. In your negotiations with toll processors, remind them that this time the farmers are in charge and negotiate a better deal than that; the industry's getting big enough that you don't have to agree to the first offer that comes your way. If you are a buyer, broker, or processor, please respect the farmer. She worked 10 months to get these plants to you.

But this haggling is distasteful to me and just further explains why I'm such a fan of vertically integrated, value-added product lines. I say this even though going the value-added route probably means, in the short term, 20 times the work and 5 times the risk. Of course, the payoff is much higher, not to mention that having a product insulates you against those wholesale market vicissitudes that follow any gold rush's initial phases.

All that said, you're also wise not to rush a product to market until it's ready. As usual, the Colville project is being smart on this front. At the same time we were in a tizzy about sulfur, Jackie and the tribe's graphic designer were putting together a package I helped design for a roasted whole-seed superfood product.

I'm biased, but I think the resulting package is super lit: The Colville Tribe's wolf symbol is on the front of the package, alongside the brand name TRIBAL HEMP and the word SUPERFOOD. The back label, in addition to some hemp nutritive facts, mentions that this is a 100 percent Native-grown product, and notes that the Persians call hemp *king seed*. All Jackie was looking to add in the early phases of the design was a historical element linking the tribe with hemp, or at least with regenerative agriculture. There were lots of traditional salmon fishing sketches and photos, but these didn't totally fit.

At about that time, Marc Grignon, a member of the Menominee Nation in Wisconsin and director of the Hempstead Project Heart native hemp organization, dragged me to the National Agricultural Library in Maryland, part of the national library system that includes the Library of Congress.[7] The staff there was energized by our hemp requests, and nearly everything we turned up, from early hemp-fiber decorticator brochures to 19th-century letters from senior USDA officials urging hemp production, was fascinating. At one point in the day, Marc passed me a weathered index card and said in his typically understated deadpan, "This one might interest you."

It was a report about a 1901 hemp planting in Colville. And so the product's back label could shout that the tribe was cultivating hemp again after a century-long break.

Jackie, for a number of solid reasons, wound up starting by wholesaling the tribe's initial marketable seed crops. Having overseen so many successful farm projects for Colville, she has been exhibiting admirable patience with the hemp project.

"We want to do a value-added product, but we want to do it right," she told me recently. "In 2019 we're growing for CBD."

I replied that this sounded like a smart move, but that I hoped she will still lead the project to a value-added superfood product as at least part of its strategy. Indian Country, like every place, needs the healthy food renaissance for which hemp is proving the vanguard. And the bottom-line payoff has real potential: Instead of less than a dollar a pound for its seed wholesale, the project would be making six dollars on a three-ounce product. Granted, that's after a lot of work and probably a $50,000 investment.

I still have original sketches for the "Tribal Hemp" product taped on my office wall: I can hardly think of something tastier for munchies-on-the-go than roasted hempseeds. It's exactly the kind of "beyond the gold rush," totally distinct offering that would earn the Dolly Parton "different" stamp. Add to that the tribal branding, and, man, I could see it in every grocery store in the Pacific Northwest. Asian markets love Native American–branded products too.

The reason for thinking about these product decisions during mid-season is that there are dozens of threads to pull together as you consider marketing even the simplest of products. I remember calling the folks who regulate commercial kitchens in Washington and learning all kinds of random bits of knowledge about approved types of sinks and wall angles. One wonders how humans survived so long before regulations.

For some reason, all these points of veteran midseason focus don't happen by folding your arms and blinking. They require scores of hours of work—the off-the-field, and thus less-fun kind. Many first-year enterprises (but few third-year enterprises) find themselves scrambling to take care of all these to-do items on top of the nail-biter of harvesttime.

Go ahead and worry about a few midseason leaf-cutter bites on your leaves or that deer that just gave birth in your field, if you're one of those people who feels better when feeling worse. But if you do manage to focus on the more distant horizon of final product, you might find you're not losing your religion over the field itself. You'll be too busy making sure the hemp survives the day you harvest it.

Have a Five-Year Plan
Colorado and New Mexico

Professional hemp work, like all entrepreneurialism, is a long, hard slog. Perhaps the polar opposite of a get-rich-quick scheme. You've simply got to budget for a five-year road to positive revenue flow. What I've noticed in the years I've been observing the industry is that folks either understand that going in, or they quit after one or two seasons.

"We could've given up after we got bunk seed in year one, or when the crop wrongly tested hot in year two," Aaron Rydell of Salt Creek Hemp tells me fairly often. "It wasn't just a strain on our business. It was a strain on our marriage. But Salt Creek is doing all right now [at the end of year three]."

I asked him what *all right* meant.

"Five figures in the bank, a used combine parked next to the barn, and a healthy crew," he said.

Not yet Silicon Valley executive remuneration, but I have little doubt that Salt Creek's dogged persistence and righteous intent will get them there if they want that. In fact, I believe most of us launching today in the independent regenerative niche have a strong shot at thriving, if we stick with it, own our mistakes, and pay attention to a marketplace evolving at lightspeed. And, of course, if nature smiles on our fields.

Despite the hurdles, despite the temptations of the CBD craze, despite the coming regulations and, yes, despite the crooked middlemen, many of us early participants remain bullish about the regenerative side of the hemp industry's long-term future because of a very strong card we hold: the value packed into each acre of hemp.

But it will usually take several seasons of work. That's why we've been generally sidestepping discussion of the gold rushers who are all in for maximum short-term profit. In a not untypical case, I recently got an email from a fellow I'd met at a conference who said his group wants to buy "a million pounds" of flower from me and my partners—roughly the equivalent of the state of Oregon's entire harvest in 2018.

"When someone throws that large a number at you, you might as well save time and delete the message," said Kentucky's Joe Hickey of Halcyon Holdings, now a permitted hemp cultivator of hundreds of acres. Joe was one of the fellows alongside Woody Harrelson when the latter was arrested for planting four hempseeds back in 1996. He added, "I suppose it wouldn't hurt to ask him to deposit half payment for the order, but you'll never hear from him again."

That's how it is in most every race—the horse that bolts fastest out of the gate isn't often the one who crosses the finish line in the money. Joe and thousands more of us are, if not yet seasoned veterans, at least beyond the honeymoon phase. We're the ones who no longer skip around gushing about hemp's "25,000 uses," but rather the ones trying to do better than break even in a righteous way on *one*.

If you're wise, you'll go in aware that your uncle's back 20 is not a check waiting to be cashed; it's a long-term project. If you think you'll make a killing off your first crop, you might as well save your seed money and buy lottery tickets. Farming and marketing are full-time jobs, day in and day out.

Here on the ranch, for example, my writing day has once again been interrupted by an actual hemp matter. This evening's issue had me, until a minute ago, calculating the number of fluid ounces in a gallon, as I plan the next pressing of Hemp in Hemp.

How many gallons, factored on our three-ounce product bottles, can I produce while keeping it both farm-to-table and 100 percent organic? Both are key components of the real-world playbook if one aims to be a regenerative entrepreneur.

The Salt Creek Hemp team knows this. Poster children of the farmer-entrepreneur model in action, Aaron and wife Margaret MacKenzie even opened a hemp storefront in tiny, economically reeling Collbran, Colorado, just to try to jump-start the local economy with the multiplier effect. What they want from their 11 acres just as much as dough is to enlist their fellow struggling cattle farming neighbors in a regional hemp-processing network. They're hoping to incite a mutually beneficial collaboration that takes their valley's economy to the next level.

"We've harvested at or below cost for our neighbors just to get them fired up about hemp," Aaron said. "It's a long road."

Nonstop work for years is a nonissue for the kind of hemp entrepreneurs who don't throw in the towel after the inevitable first-season wake-up call. The endgame is just, ya know, the revival of rural America. The rebirth of Main Street. The quality all initial modern hemp players I've met share is indefatigability. The trait they all lack is whininess.

Non-whining beginners are welcome to hemp, because we're all beginners. A third-generation Kentucky tobacco farmer, Kendal Clark, in softening his criticism of the performance of some Canadian hempseed he had planted in 2016, said, "Well, heck, they've only been at it twenty years up there." Up to that point, I had thought the Canadians were the wizened pros at the crop. But Kendal's tobacco genetics go back a century and a half.

With very few exceptions, we've all just been called up to the majors. And this is why I'm sympathetic to all those exuberant newcomers who want to see what happens when they throw an acre or 20 of hemp in the ground, as opposed to the school of seasoned farmers who sometimes espouse the "Man, what are these greenhorns thinking, planting without developing products and markets first?" viewpoint. The way I see it, we've all got to start somewhere. Yes, having a post-harvest game plan is vital. But you don't know squat until you've actually tried to grow this crop.

So plant away, people. Only 1 percent of Americans are farming today, as opposed to 90 percent in George Washington's day, and 30 percent when cannabis/hemp prohibition began in 1937. A return to a 30 percent farming society is a viable goal.

This is not a flash in the pan. There is real demand for our products. When *Hemp Bound* came out, in 2014, there were zero federally legal hemp acres planted in the United States. In 2017, there were 23,343 acres. In 2018, 78,176. Unsurprisingly, projections for 2019 acreage are for 175,000 acres, according to Eric Steenstra, executive director of the advocacy group Vote Hemp.

The emerging wider biomaterials economy really does have the potential to transform the worldwide industrial pipeline into a regenerative mode. Visualize compostable cell phones with both their outer plastic casings and their interior battery components made from your and my favorite crop. The latest versions will be delivered in electric vehicles powered by hemp batteries that are charged by the sun.

Yet hemp is still such a toddler, it can hardly even be considered a niche crop by big-picture agronomists. Those 78,176 acres in 2018? That's compared to *89.1 million acres of corn*, according to USDA figures. We've got a long way to go. Although corn acreage (essentially meaning GMO corn, doing little good to farmer, consumer, or soil) was down 1 percent in 2018. Domestic hemp acreage grew 334 percent. So the trend is in our direction. Only 89 million acres to go. Or 234 million if we set the bar at overtaking corn, soy, cotton, and wheat acreage. But the shift is on. And we're gaining fast.

———————

Let's say we do rally, collectively, as a mass movement of independent hemp entrepreneurs, in this bottom of the ninth. What does wider world victory look like? How do these regenerative values that hemp is about to popularize strengthen communities and rebuild Main Street?

One of the best real-world models I've seen in action is right in my home region, where 32-year-old Nick Prince has morphed his successful computer repair business into a nonprofit called Future Forge. The venture scooped up 13,000 square feet of abandoned warehouse space in what had been a ghost town corner of our nearest town, Silver City.

The space, formerly the metal and machine shop for our region's copper mining conglomerate, is full of terrifying industrial-grade equipment that can do everything from manufacture aircraft components to dismember a guy like me in under a second. Prince and his team have supplemented the analog hardware with sewing machines and a rack of 3-D printers.

Future Forge's plan? "Reclaim the local economy and rebuild Main Street, by producing everything possible here from renewable materials," Prince told me recently, as we toured the three-football-field-sized facilities. Though it has a gorgeous, Wild West–era Main Street and an organic ice cream truck, Silver City needs this radical localization; downtown businesses tend to have high turnover, and New Mexico's poverty rate in 2017 was close to 20 percent.[8]

At its core, Future Forge is geek and maker central for my ecosystem. But agriculture is deeply embedded in the plan. When I told him I was about to cultivate hemp in New Mexico for the first time, Prince led me straight to the 3-D printers, pointed vigorously, and said, "What do you think we should make from the first harvest?"

I thought for a moment, remembered the hemp plastic goat I'd printed for my kids in Colorado a few months earlier, and said, "Compostable take-out servingware."

You can access all of Future Forge's equipment for $60 per month. My family has joined. "As the tools democratize, the future localizes," Prince said. Farmers and digerati as soul brothers, on the same mission, for the same reasons. We were already plotting the end of disposable plastics in our county. And so the late-inning rally begins.

As I left Future Forge headquarters that day, I found myself thinking about comedian Kevin Hart's bit where he describes being surprised, upon achieving success, to find himself living not among other actors and comedians, but among dentists and lawyers. Therefore he's nudging his kids to become dentists or lawyers, so they, too, can live in froofy neighborhoods.

To "dentists and lawyers" I'd like to add "farmers and healthy-product makers." There is no Dentist Aid concert. Much as I adore the music, it'd be fantastic not to need Farm Aid. So as you glean the hopefully practical tips embedded in this book's escapades, I hope you also keep this desired result in mind: Living in a world where parents tell their kids, "Work hard and someday you might grow up to be a farmer."

CHAPTER NINE

Combining Killed
the Radio Star

Oregon and Vermont, 2018

The secret is to bang the rocks together, guys.

—Douglas Adams,
The Hitchhiker's Guide to the Galaxy

I f there was one stretch of the hemp calendar that I thought I had nailed down as a veteran of a dozen harvests, it was "when a seed crop is ready." I could recite the mantra in my sleep: Harvest when 70 percent of the seed is ripe (meaning brown, no longer green), is emerging from its calyx (leafy seed cocoon), and can't be crushed by squeezing between your fingers.

I'd lived by these few rules without fail in my own fields, taught them to friends, partners, and research colleagues, and wrote about them in my journalism. Then 2018 happened.

When Your Crop Is on
Its Own Harvest Calendar

As usual, the Samurai was taking a long time to mature, on both coasts. I'd call the cultivar's long season one of its few flaws. Its seed-to-harvest cycle is 120 days, easy. Some cultivars are done at closer to 100. We're working on this in our breeding efforts.

Still, late-maturing seed had been a previous year's midseason panic. I knew to be patient because the payoff was worth it. "We will sell no hemp before its time" and similar platitudes. For whatever reason, though (fires? latish planting?), this season the Oregon crop was taking even longer. Halloween was approaching—Margaret was already putting together costumes for our planned October 31 harvest.

When I dipped into the Emerald Triangle on October 28, outrunning seeping vog from Kilauea, which was erupting while I was working on a Hawaii harvest, the seeds were maybe 50 percent ripe across our three-acre field. I didn't want to believe my eyes. But when I checked the same flower clusters each successive morning leading to harvest, impressively seed-dense though they were, it looked as though they were ripening maybe 3 percent per day.

Much as I loved the tasty seed sampling these dewy strolls included, I did not enjoy the feeling of squishing so many green ones mere days before harvest. There were hundreds of thousands of seeds packing the sea of banana-sized flower clusters, but nearly half weren't fully brown or emerging from their calyx. The hemp decided it was on its own schedule, not mine.

On the other side of North America, Cary, Colin, and Erin surprised me by reporting the same 50 to 60 percent ripeness in Vermont. In that East Coast field, snow and hard frost had come and gone twice already. And while the hemp was proving as hardy as ever—mostly not noticing the weather, and recovering immediately where bowed by intense wind—Vermont was my destination immediately following the Oregon harvest. Regardless of the seed-ripeness percentage, we had no choice but to harvest. It was time.

That had become my routine in 2018: First learn how to do something from Edgar and Margaret, then hop on a plane or into the camper and do it again with the Vermont team. Worked at planting. But what I was proposing for harvest was a plan B, to say the least.

The reason you aim for 70 percent seed ripeness at harvest is this: Whether you harvest by combine, oxen, or hand, you're going to shake some overripe seeds loose in the harvesting process. Consider it bird prenatal care. To my mind, it's also in sync with the biblical mandate to leave the corners of your field for anyone who may need the self-serve harvest. The resulting feeling of moral righteousness slightly reduces the stress of parting with valuable seed.

When you hit that 70 percent seed ripeness, clever humans have discovered over the millennia, your loss will be at a minimum. You'll have some seed too ripe and some not ripe enough, but you'll maximize your seed harvest, like Goldilocks' just-right porridge.

Since we've seen how I reacted to a text about a few yellow leaves, you can imagine my dreamscape those first nights in Oregon after another day spent walking a crop that was not ready for harvest by the definition of *ready* that I knew and trusted (other than its enchanting fragrance). It looked and felt like harvesttime to every plant and animal in southern Oregon except our hemp crop. But Edgar was unperturbed.

"That won't be a lockjam," he said. "Crop's close enough to finished that the seeds'll ripen to seventy percent while we dry 'em. At least seventy. You'll see—that seed-drying process is like an energy Gore-Tex. It finishes the crop for us. There won't be any gridlog."

By this point I suspect most readers will be unsurprised to learn that my hemp professor was completely correct. We got the harvest drying, and 3 days later the seed crop was 70 percent ripe, maybe 80. Seeds were too strong to crush. The harvest was a success. The crop a healthy and a valuable one.

Every hemp harvest is a lesson or two. Oregon's in 2018 were (1) social media loves a pic of hemp farmers harvesting in Bernie Sanders masks, and (2) as long as you get your seed drying immediately, you'll be okay if you harvest a bit wet and green at 60 percent seed ripeness. It's not plan A, but then what in hemp is? What in life is?

Contemplating a Hand Harvest

While the seed-ripening issue was being dealt with, 2018 also became the year I found a way to make an efficiency argument for my preferred mode of harvesting. Namely, using human hands, perhaps aided in large-acreage fields by some of our stronger nonhuman friends, like pack horses and oxen. The question I'd been looking into was, can people do as good a job as John Deere?

Just so there are no secrets about my motives, the reasons I prefer hand harvesting to diesel-combine harvesting, in order, are (1) it's quieter (I like to hear bees), (2) it uses less petroleum, and (3) it allows more time for berry picking.[1]

Since on both coasts the Samurai crops were in the three-acre range, we had a genuine choice to make. Three acres is a lot of ground: 130,000 square feet. And yet, not so much. From a high vantage point, you can see your entire field in your frame of vision.

So we weighed the pros and cons. To hand-harvest or fire up the combine? That was the question, for several weeks.

In Vermont, where there are about six combines in the whole state, we always had Charlie waiting in the wings, meditating over a brew in his Yankee-farmer retro-chic garb. We'd call the backwoods genius should we cave to the combine's alleged efficiency. He'd find a machine. He'd make it work. But you can summon a Charlie only so often, and I wanted to save him for a forthcoming solar jet pack idea.

In Oregon, too, we'd been pricing combine rentals in the weeks leading to harvest. The Emerald Triangle, that mountainous region known for high-value specialty crops like grapes, cannabis, and Shakespeare festival tourists, didn't have many either. Plus Edgar and I acutely remembered our adventures in seed drill malfunction five months earlier. The industry was going so acutely ballistic in Oregon by 2018 that even finding farming equipment available for rent, let alone cleaned before returning, was becoming a challenge. It was a seller's market, from wheelbarrows to farmland.

Oregon had 230 hemp applicants in 2017. (Edgar holds permit #1 from 2015.) The Beaver State had 560 in 2018. There were only 3,546 permits issued in the United States that year. Impressive, yes, but all these numbers will look nostalgically embryonic in coming years. By 2019 Oregon had 1,500 farmers planting 50,000 acres.

In the end, we decided to hand-harvest in Oregon that Halloween. Nothing like grandchildren and cousins to beef up a work crew. Edgar's brother-in-law Mikey brought a crossbow with him, in case venison was to be our All Hallows' Eve main course. We wound up shooting at a hay-bale target we labeled with a sign reading PROHIBITION.

Turns out hand-harvesting, despite hemp's famously long taproots and Samurai's 10-foot average height, is pretty easy. Especially when you plant tight, as Edgar somehow knew to advise that year.

"I got an idea on spacing," he'd told me back in the spring, when suggesting we plant about twice the per-acre poundage that I would have predicted. He started heading to the center of the field. "Come on out here and tell me what you say."

I'd previously planted Samurai ganjalike, on those 15- and 30-inch bases, harvesting the resulting hemp redwoods. These had defeated both our combine and our arms, requiring machetes in some sections. But Edgar was proposing seven-inch plant and seven-inch row spacings for 2018.

"We might get a little less flower and seed per plant," he said. "But I bet it'll be close and it'll make our life a heck of a lot easier come harvest."

"That sounds logistical," I said, suggesting we go for it, in the name of agronomic research.

Covering the field 120 days later with three teams of two people (plus dogs and bees), I am so glad we did. It took us 2½ very pleasant days to harvest the Oregon field. Each human harvesting machine bent one stalk at a time down over a one-ton, chest-high grain bag. Once a stalk was arced gracefully over a bag, the farmer shucked off the seed and flower with a single, sliding motion.

Just like that, the seed and flower harvest was done for that plant—only 180,000 to go. That left the fiber. When ready to pull the stalk from the soil, the farmer assumed a sort of wide straddle stance. What followed was a medium-strength yank (lifting from the legs, of course). Thanks to the thinner plants, hand-pulling fiber proved surprisingly doable. Perhaps 1 in 10 plants required an extra pull or grunt. Suddenly, upon shaking off all the luscious, mocha, highly organic aggregate stuck to the roots, we had another stalk of top-shelf, textile-grade fiber to add to the growing piles.

Time for another mouthful of blackberries. These breaks ended when, chin dripping with juice, you noticed another team's seed bag, a few rows over, was getting way ahead of yours. This was a half dozen people's waking lives, in 5- and 10-hour stretches, for several days.

It all looked very Halloween-y as the field cleared and the fiber piles grew. The Emerald Forest was spotting and rusting, ready for a winter coat and a rest. So was I. I noticed I was engaging entirely unfamiliar, primeval muscle groups, completely distinct from the planting-time muscles.

"This is farming," Edgar said at sunrise on November 1 as we hobbled back to the field for the second day of harvest. "Work until you drop. Get up and do it again. Repeat until dead."

I still pick farming over anything indoors. We wore our Halloween costumes again on day two, since they were already torn and covered with bits of leaves glued on by trichomes. Edgar dressed as Bernie. I was, somewhat predictably, a space cadet.

Combine-Oxen Face-Off

It all went so smoothly, quickly, and, most pleasing of all, quietly in Oregon in 2018 that I decided to do a little bit of research on farmers' historical experiences with pre–John Deere modes of harvest.

I started with the Founding Fathers. George Washington loved his oxen. In his diary entry of October 15, 1787, Washington notes that he had eight available for harvesting on one of his farms. As soon as spring sprung the next April, he was out with them again, recording a successful planting of barley and grass seeds with a harrowed oxen.[2]

Two hundred thirty-one years later, I had the great honor of harvesting hemp in those same fields with an 18th-century sickle: Dean Norton, Mount Vernon's terrific estate horticulturalist of the past 50 years, invited me, Virginia hemp activist and entrepreneur Jason Amatucci, and a few others to the first hemp harvest on Washington's farm in 200 years. And that same day, I had the opportunity to thank a team of the red oxen who had sown the hemp fields in the spring. With no spraying. I'm seriously considering pack animals when my personal fields scale up. Can't be harder than goats, I thought as I left Mount Vernon and checked into National Airport with terpene-reeking fingers fusing to my ID.

While the plane taxied, I wondered what first motivated farmers toward diesel combines. Maybe it was effective advertising, maybe it was a sense that the grass is greener: Your neighbor invested in this soot-belching miracle machine. Maybe it'll make the life of the farmer easier. Heaven knows, we need that still.

As I watched the Washington Monument, the Capitol, and the Mount Vernon hemp field shrink to miniature, it occurred to me to examine whether anyone had researched the relative efficiency of mechanized versus nonmechanized harvesting. Surely folks before myself had stood around long enough holding the wrong-sized socket wrench during a FLOATER delay to wonder.

Indeed, they had. Just as pack horses and oxen were going away in the United States, in 1935, *The American Economic Review* conducted a study called "Tractor Versus Horse as a Source of Farmpower." It concluded that a tractor can, per hour, do more work than a pack animal. Can. When it's working.

A 2006 Nebraska field test between a 1936 tractor and a team of horses showed the tractor, if working nonstop, was about twice as fast at plowing a 40-acre field. Which isn't that much. And it didn't cover FLOATER constants such as breakdowns, fuel and maintenance costs, and hearing damage.

When it comes to cost, the pack animal harvest wins by sevenfold, according to a 2015 article in *Small Farmer's Journal*. "A team of horses (including food, bedding and equipment)," the article noted, "averaged $3.39 per hour to operate, versus $21.21 for a tractor."[3]

Plus you get to crow about "petroleum-free harvesting" on your labels and social media. To my mind, it kind of comes down to your preferences for exhaust: burning dinosaurs or fetid manure. And to your preferences for decibel level: deafening or idyllic. My alarm clock on the Funky Butte Ranch at this time of year (peak spring) is the local hummingbirds finding the Russian sage outside the bedroom window at sunrise. You can smell the flower beds from inside.

By contrast, the first and only time I heard my 2016 farm partner John Williamson swear was because of a different kind of smell. I had just broken for lunch after a morning spent hand-collecting the flower expelled by the vegetable oil–powered grain combine. I was in the kitchen with Gram, John's in-law and our cook that year, eating homegrown potatoes, when my phone sang and it was the normally taciturn farmer screaming from the field, "The f——ing blades are turning blue!"

"That's not good?" I asked.

"You should smell them. They're melting. If I hadn't checked when I did, it would have burst into flame."

Williamson, remember, is a classic non-whiner. After a half century of farming, he was seeing something new.

Forevermore at every combine harvest, I make sure everyone periodically hops in front of the machine's header (the front end combine attachment sporting the actual blades) with machete or *Crocodile Dundee* knife drawn. The goal is to detangle and remove the latest spooling layers of (trust me) one of Earth's strongest fibers from the blades buried beneath. Here's a survival tip when executing this maneuver: *Shut off the engine first*.

Hemp-related combine fires are almost always due to fiber wrapping: Even without the blades turning blue, FLOATER demands that nearly every hemp combine harvest is going to come to a halt every couple of acres in order to hack off the stalks from a knotted weave around the blades. These can wrap so densely that sometimes it seems to me it would be easier just to bring out some seamstresses to the field and let them loom it.

Melting blades are just one combine issue. Then there is the need for continual blade sharpening. This unavoidable task alone sent me shopping for oxen on Craigslist. Here again, John is the model. He spent the dawn hour on harvest day in 2016, as he had on planting day, in his shop.

My job that morning was to carry in the aptly named, very heavy "star blades" from the '86 John Deere combine to John one at a time, while he sent a curtain of steel flakes hailing into my hemp shake. I'm hard-pressed to think of anything I'd less like to have as part of my work life than combine blade sharpening.

Still, even with the constant stream of combine horror stories in the hemp farming world, the arguments I put forth here are not black and white. My foil in the debate is good ol' Aaron Rydell of Salt Creek Hemp. His tax return occupation should read *grease monkey*.

For years before coming to hemp, he fixed choppers. Aaron likes noise. Reminds him he's awake. Or else, he can't really hear it anymore. Diesel, to him, smells like victory. Since getting Salt Creek's long-out-of-warranty combine supercheap, Aaron, forty-four, has enjoyed nothing more than a loud morning spent welding. This allows him to entertain people like me with tales of jury-rigging and tricking-out. Aaron and I agree on just about everything except RPMs and pace.

When I start talking about the efficiency of quieting things down, he always reminds me in the perpetual shout his ears demand, "Yeah, but you only have to feed your diesel engine when you're using it. You have to pay employees and feed oxen year-round."

And, I would add on the animal husbandry front, know how to train them, and keep them healthy and happy. Plus build and repair their harnesses and yokes and such. All, to me, much more pleasant and intuitive tasks than adjusting manifolds and lubricating drive shafts. If Aaron rubbed a magic lamp, by contrast, he'd ask for a Sunday spent hoping a two-ton chassis doesn't roll on him as he wriggles a belt wrench around a muddy fuel filter while blowing helplessly at horseflies.

There is no AAA for John Deere. Not when you're out of warranty. Still, at some acreage threshold (mine might be higher than some folks'), combines come into play. While only two of Vermont's 2018 hemp farms were larger than 10 acres, in the heartland, when you say your farm is "ten," you mean 6,400 acres. As in 10 "sections" of 640 acres each. This goes back to the Homestead Act, and is how farmers think in the Nebraskas and North Dakotas of the world. Combines are part of the family in this ecosystem. People raise them like chickens.

The basic idea of the grain-harvesting combine is this: As you very slowly move across your rows, the plants, some of them 12 feet or taller, are cut, then sucked by the dozens, into the combine, along the header's revolving reel. The action of the reel, aided by an auger, agitates the stalks, removing and depositing the ripe seed in a catchment bin. The deseeded flower and stalk then shoot out the back of the combine.

Combines range from 12 to 40 feet across the header, cost six figures new (seven if you want the bots that do the driving for you via GPS), and are very dangerous to be around.[4] Make sure your fingers, dogs, and children are clear when the deafening monstrosity starts up: You don't want any of these to get sucked into the gaping jaws of their blades—the things are almost strong enough to penetrate hemp fiber.

I've noticed that experts position the header a few inches above the ground. Too much clearance will leave Viet Cong–like fiber spikes in your field, which are difficult to plow under or otherwise deal with next season. However, if you position the header too low, your blades can smash into the earth, hyperextending them and ending your harvest until the new John Deere models come out.

Once you've got your seeds inside the belly of the combine, emit a profound sigh of relief. The seed part of the process from here on out becomes much easier, if your drying setup is on-site. Like Fred Flintstone on his comparably sized dinosaur, all you have to do now is simply drive the combine to your silo, deposit the seeds through a sort of reverse Shop-Vac on the combine, and run the giant fan that you set up midseason below the grated floor of the silo.

The harvest at this point smells scrumptious—that's the terpenes again. Now's when you pull out that seed-moisture tester. When the seed is down to 8 percent moisture, probably after a day, maybe two, transfer it from the silo to your seed cleaner.

My combine mentor is North Dakota's Roger Gussiaas. He's the kind of guy who suggests we grab a six-pack and hop on the ATV with the dogs to check out the back 10,000. In 2016, confused state and federal officials made the significant error of wrongly hassling the permitted Roger. Since then, this baby-faced, religious fellow has proved himself North America's biggest organic hempseed oil producer; his Healthy Oilseeds plant is running 24-7 and has been expanding nonstop since we met.

Before he came to hemp, Roger harvested another strong fiber crop, flax, for two decades. So this guy has the ultimate combine creds. My invariable practice today if an engine is imperceptibly leaking or a robust fiber crop has the header in a chokehold is to call Carrington, North Dakota, the way the Apollo 13 crew called Houston.

For the most part, Roger is a soothing voice in the harvest discussion. He's of the opinion that much of hemp's reputation for harvesting difficulty is a matter of technique.

"To help cope with hemp's long fiber, open your header wider than you normally would, and come at it from a less aggressive angle with your blades," he told me. "Even when your seed is ripe, you're harvesting when the fiber's still a little green, which makes hemp an unusual crop. That's why my advice is, be gentle on the plants. Let them float in. Rotate your header at a slower speed and approach at a concave angle. If you're at a sharper angle, you'll get less cutting and more tearing."

I don't know how much this advice has helped me when confronted with a combine harvest. Whenever I'm inside a combine cab, I become the toddler who first got to sit in Dad's '77 Buick and clutch the steering wheel. Both times (Buick and combine) I recognized that there was much more to this than I had imagined. I'm pretty sure a 737 cockpit doesn't have more buttons on its panel than a 2018 John Deere.

But it's the levers that are really scary. They stand like Easter Island statues near the throttle. Shift the wrong one, and you'll lower an 800-pound steel header on your partner, when you intended to shift into reverse. The whole combine harvest day is a potential Three Stooges episode.

My real concern with combining hemp is the continual fiber-unwrapping parties. Even Roger acknowledges this symptom of FLOATER: "Oh yeah, you have to stop a lot during a hemp harvest," he said. "You budget for it."

And even when you do everything right, a combine can burst into flames. Aaron understands this firsthand. In 2017, the up-to-that-point smooth Salt Creek harvest was barely saved thanks to a heroic half-mile round-trip sprint by longtime business partner Joe Koeller to retrieve the small fire extinguisher that was supposed to be residing in the combine for just this reason. It was an effort that left our protagonist returning his breakfast to the soil from which it came.

Turned out not to matter, what with the hot test a few weeks later. But whenever he hears me tell the "blue blade" story from Vermont, Aaron says, "Oh, so the combine didn't actually catch on fire? How lucky can you get?"

Professional Hand Harvesters

The combine fire of 2017 is still a sensitive topic for Aaron. As Joe tells it, when a bearing in the chaff chute broke (this is the chamber that expels the part of the harvest you don't want in your seed hopper), the resulting friction set the combine's exhaust alight.

Mechanic shoptalk aside, Joe's point was, there is no way to plan for or prevent these kinds of fires. "We did all our maintenance," he said. "Changed the belts. These things just happen. Hemp fiber is strong. Bearings wear out."

This fiber-harvest hassle is why even Grant Dyck, an early Canadian hemp farmer I profiled in *Hemp Bound*, told me he quit hemp.

"Okay, I guess under ten or twenty acres it could make sense in some circumstances to stay medieval on it," Aaron conceded recently, which I know was hard for him to do. So I hereby say, "Ha ha. Told ya."

Kidding aside, hemp's reemergence in the digital age is showing, for the first time in generations, that a farming family can make a living while hand-harvesting on small acreage. At least on paper. But is anyone yet living the regenerative model and actually earning real income?

Monoculture is big business—$2.3 trillion worldwide. Right from the first Farm Bill provision's passage in 2014, I (and others) recognized that Big Ag and Pharma had their eyes on the box store CBD market. It wasn't hard to see—all you had to do was look at who was sponsoring hemp conferences and who was working toward bogus patents on parts of our ancient plant.

In an effort to help farmers leapfrog this side of the industry, I started proposing these value-added strategies to independent regenerative entrepreneurs in 2016. After a convention keynote or college lecture back then, I'd sometimes wake up feeling that cheerleading this ambition might be a bit grandiose at best, if not reckless. We were talking about changing the planetary farming MO of more than a century. Were we actually suggesting that a challenge to business as usual would come from perhaps the most neglected profession of the past century, independent farming?

Then, in early 2018, not one, but two Wisconsin farmers returned to a Midwest hemp conference (held in a—ho hum—blizzard-induced state of emergency) to tell me that my words had motivated them at the conference the previous year, and that they had both successfully harvested *and* made some money in their debut season. And both did it with hand harvests. Making bank in year one? That was more than I could have said three years earlier.

"We're keeping the lights on," 56-year-old Steve Tomlins told me of his initial five and a quarter acres of organic hemp outside of Delavan, Wisconsin. "We sold a hundred fifty pounds of flower to a processor, and market the rest as high-end smokable flower."

That's from a one-quarter acre of sinsemilla clones. The remaining five acres was a grain crop for which Tomlins said the target market is farmers and processors in nearby Midwestern states like Illinois as they come online with hemp. He and plenty of others since have demonstrated that if a lot goes right and you work your tail off, you can reasonably expect to make a living while building soil and your local economy.

"My advice to folks coming in," advised Tomlins, echoing Berry. "Is to say no to wholesalers offering bad prices."

Independent agricultural entrepreneurialism is no longer a fantasy, which is a relief. While I won't be surprised if the efforts of early-phase enterprises prove to be nearly as risky as all entrepreneurialism, stories like Tomlins's have buoyed my confidence. So much so that in North Dakota later that year, while we wore hair nets and watched organic seed oil drip down into storage units from one of the huge Healthy Oilseed presses, I broached this topic with Roger Gussiaas.

"Does a big-time, multisection farmer like you think this talk of hand harvesting and pack animals is crazy?"

"I don't think it's crazy for a second," he told me. "It's a legitimate discussion. It's a question of scale, time, and available hands."

"You can hand-harvest and still be a pro?" I asked. "If you go value-added?"

"Let me put it to you this way," Roger said. "I've farmed five-thousand-acre crops, and my fifteen acres of hemp gave me more problems, what with the fiber wrapping. I'd say you have to be a pro as a hemp farmer, regardless of the acreage."

Whether one can make the case that even larger harvests might be handled equally efficiently without diesel engines depends on how many *Homo sapiens* (or trained oxen or workhorses) are available to work with single-minded focus for a couple of days at a stretch.[5] And when farm mechanization moves to solar-powered electric, then I'll get interested in it. I mean it. If we can automate in a way that's both regenerative and, equally vital, quiet enough to hear bees, I'll learn my ohms and my volts. I'll even get myself a welding mask.

It's already starting to happen for other crops. The farmers at Hana Ranch, a 3,600-head cattle operation and orchard on Maui, converted a 1948 Allis-Chalmers Model G tractor to electric power because, in the words of ranch mechanic Evan Fourtenout, "there is no maintenance required, except for two shots of grease once a year to keep the main bearings lubricated."

Writing in the ranch's blog, Fourtenout also said tractor performance is superior on electric power. "The electric motor eliminated the need for a clutch, so when the throttle is opened, the motor puts out almost 100 percent of its torque throughout the speed range." That's useful for the slow speeds you employ with a tractor.

And, he added, "There is almost zero noise and actually zero emissions. Consumption and waste are things of the past as well, as we never have to change the oil or fill the tractor up with fuel."[6] Hard to beat that as a company marketing storyline. Now I'd like to see the first electric combines.

The larger point of all this, at least while we still live in the noisy combustion era, is that when you have an Aaron, Roger, or John Williamson on hand, you can deal with engine maintenance issues and minimize your FLOATER downtime. If you are not a diesel mechanic and don't aspire to be one, don't have one on your hemp team, and can't afford to hire one for two full days in your remote fields, here's some encouraging news: A hand-harvest brings in the same amount of hemp.

The 2018 Emerald Triangle hand-harvest gave us two tons of flower and about that much in textile-grade fiber, some of it that magical purple phenotype. I recognized the sensation I felt, as I peeled off my costume that night in Margaret and Edgar's basement guest house / processing room: It was that deep feeling of satisfaction that I'm programmed to feel as a human who has provided for winter sustenance. The harvest is in. We shall eat. It's when I feel closest to being a squirrel.

It explains why, at every harvest with my kids, I wait for us all to have an armload of plants. As we deposit them in a silo or bag, I say, "There, we did it. Another three pounds of flower, ten pounds of seed, to feed our family."[7]

It's also why, 4 days later and 3,000 miles east, I made the hand-harvesting pitch to the Green Mountain squad. It took minimal convincing. All I had to say was "If it proves hard, waffles are on me."

Boom. Another smooth, fast harvest. Second in a week. Maybe even easier than the Oregon harvest. Part of that was payoff from an effective planting: Charlie's antique seed drill had provided—by a solid margin—the most evenly spaced rows I'd ever worked. That 2018 Vermont field looked like it had been planted by robots. Made it so pleasant to harvest.

Like an AP student placing out of a college prerequisite, we became exempt from FLOATER rules when we unshackled ourselves from John Deere. Of course, the exact duration of mechanical delays were technically replaced by berry-picking breaks. But I would argue that those delays compose a separate category from FLOATER delays because of their energizing nutritive bursts.

Just as we were working our final pass, with a quarter acre to go and chilly hands, we all wordlessly, simultaneously, looked up and realized it was snowing. Large wet flakes in paper chain shapes were blanketing our fiber piles and our hair like confetti. Our faces burst into smiles inside our three-acre snow globe.

No need to help, we all radiated as we became invisible in the whiteout. *We're just lost and blissed out in a hemp field again.*

The November snow was still on the ground when we processed the crop into product in March.

Dealing with Hand-Harvested Flower

Since we've now covered the mechanics of seed and fiber harvesting, what about bringing home the dioecious flower that puts the *tri* in tri-cropping? The flower, of course, is where the eminently marketable cannabinoids (CBD, CBG, THC, and the rest) reside.

You'll have tons of flower sitting in your 20-acre field or in bags now (depending on whether you harvested by combine or hand). You're wise to treat it delicately while it's fresh, in order to most effectively dry, cure, and store it prior to selling it raw or making it into your product.

Weather matters, too, when it comes to the flower side of the plant. In fact, it might not be a bad idea to do the opposite dance you did following planting to ward off rain at harvesttime: You want to avoid botrytis (also known as *bud rot*) in your valuable flower.

"If someone says they have a ton of flower, that's not enough information, even if they show you the cannabinoid profile," said Andrew Bish, who as a harvest equipment salesman sees as much just-plucked flower as anyone. "It can mean a wide range of quality. Are you taking care of those trichomes, which are delicate, or are you mashing your whole harvest together in a green pile? I can look at two harvests of the same cultivar, and they are not worth the same."

To some degree, the current wholesale pricing system for CBD flower favors churning the flower and stems into that pile of mixed green material.[8] That's because price-to-farmer (at the moment) is based on a point system where each percentage of CBD in the flower's cannabinoid test means a higher per-pound price. So if your stem adds weight, it also adds value. Plus this "batch harvesting," since the material is already chopped finely, leaves it ready to feed into large-scale processing machinery. But this is not a top-shelf harvesting mode.

What we did during the 2016 combine harvest in Vermont was simply snip the best flower branches off tens of thousands of downed plants and place them on tarps, hauling them periodically to the shop for hanging by the branches to dry and cure.

That space filled up quickly (flower-drying capacity being another checklist item to add to your midseason tasks). To deal with this once the seed harvest was dry and bagged, we tossed a few hundred pounds in the silo to see if forced air drying also worked for flower. It did. The flower came out with its trichomes looking impressively crystalline and intact, even after running through the combine. These are the flowers that went into the first run of Hemp in Hemp.

At harvesttime, hand gathering the deseeded flower emitted by the combine is a fun, sticky job. Much preferable, in my view, to that of the poor sap wearing ear protection inside the combine cab. But this mode is not exactly what you'd call efficient. Certainly not as automated as the seed component of a combine harvest. We wound up bringing in maybe 30 percent of the flower from that first large Vermont harvest. Still, we got tons of material, more than enough for the first run of Hemp in Hemp. When I find myself wishing that we had found a market for the rest, I remember that today we would have. Mainly because the Farm Bill and resulting markets allow us to more powerfully negotiate with processors inside and outside of Vermont.

For folks harvesting sinsemilla CBD crops, I recommend emulating the best outdoor cannabis farmers in the world, the Emerald Triangle farmers I profiled in a book called *Too High to Fail*. These artisans don't merely hang their flower-loaded branches in upside-down Y-formations in barns with good cross-ventilation. They get super scientific about it, measuring humidity, oxygen, and CO_2 levels during drying.

Once they get to curing, a stage almost always omitted by today's early hemp cannabinoid farmers, it really becomes an art. My favorite ganja farmers place their entire dried harvest in gallon-sized, violet-tinted glass mason jars for a month or more, "burping" the gases out according to a schedule they'd been perfecting for three generations. No Mylar bags. No nitrogen infusion. This delicate postharvest handling is something for hemp cannabinoid flower purveyors to consider, if marketing top-shelf raw flower as Steve Tomlins does. Might be for the smokable market. Might be for salads at Whole Foods.

Given that current wholesale prices for high-grade hemp flower can rival psychoactive cannabis prices, how might efficiency on the flower side of a large-acreage harvest be improved? This is where the clever folks at Bish Enterprises come in. I discovered this one dewy morning on Colorado's Western Slope during a visit to the Salt Creek Hemp ranch.

Unless they plug clones, outdoor flower-only farmers tend to plant between 1,500 and 4,200 seeds per acre. (You'll be on the higher end if planting dioecious seed, as you'll be pulling males.) While some farmers encourage those massive, branching, flower-dense plants you see in ganja calendars, Andrew Bish points out that propagating smaller plants makes harvesting much easier with equipment like his Hemp Handler 6031.

"Plus your plant is putting all its energy into flower production," he said. "You get less biomass per plant, but equal or more per field, with a more efficient harvest."

Salt Creek has used the Hemp Handler 6031, and its farmers grow small plants from clones specifically to make harvest easier.

"People laugh at our short plants," Aaron said. "But they are compact little flower machines."

A further advancement we might see in coming seasons will emerge not from the harvesting side, but the breeding and processing sides. Some breeders are developing flower cultivars that can be processed whole; an entire harvested plant is tossed into a special processor that extracts the cannabinoids and terpenes without first separating out the flowers.

Myself, I might stick with hand harvesting, and I might not. I like wearing the "lovingly harvested, top-shelf" mantle. Keeps me in the field, inhaling terpenes. But as my fields get larger, I have some decisions to make. Oxen and pack horses will be my first choice. If they work, great. But if the electric combine is online by that time, I might just fire one up.

I don't know if I'll ever be lucky enough to harvest in the snow again. I hope so. I know we all still talk about it. I remember thinking, as we whitened that evening in Vermont, that traditionally, a farmer would call this perfect timing: winter arriving just as the harvest is in. Time for mending harness.

Not for me, and not for Colin and Erin. Per Margaret's Law, we had to process our respective products. I was already trying to remember how many hundreds of empty Hemp in Hemp bottles I had stored under Cary's daughter's desk six months earlier. How many caps did I have to order? What would be the final hit to my bank account for product labels?

A few hours later, while the bulk of the harvest was drying at the popcorn farm, we brought some seed and flower home to Cary's for the long-planned waffle feast. Oh man, fresh hemp waffles soaked in home-tapped maple syrup and topped with raspberries. It sounds like an epicurean binge, but actually it was a business lunch: I suspected I might need the energy boost for the intense processing that was to come.

I was kind of right. What with homeschooling and looming book deadlines, it took four months before I got to processing in early 2019. But superfood waffles with friends at least helped me start the journey.

Plan for Plan B

Nearly everyone in the hemp industry is on to plan B, and some of us are on plan D or M. Whatever you think your objective is in hemp today—selling highly concentrated CBD tincture, making hemp ukuleles—is likely to change, at least slightly. Probably more than slightly. Don't sweat it. Heck, even something as seemingly fixed as magnetic north moves 34 miles per year. Just recognize that the plant is in charge. Half a decade into the modern game in the United States, the long-term players who are actually enduring in the industry have all had to implement the pivot that HempFlax's Reinders felt was just part of running a hemp enterprise.

It's not just you and I, the relative newcomers, who are affected by this phenomenon. The pioneers of the modern hemp industry itself in Europe, Canada, and China are on to plan B, too. Take John Roulac, founder of California-based Nutiva, which crossed $100 million in annual sales in 2015. Even he has decided to pivot. The former redwood-sitter's ubiquitous hempseed oil provided a key source of pre- and postnatal protein to my growing family before we began cultivating hemp ourselves. But after a quarter century working exclusively on the seed side of the plant, Roulac launched a CBD company called RE Botanicals in 2018.

The seed and fiber focus was so hard won—including a 2004 court victory over the DEA that allowed even nutritive hempseed products to be marketed—so ingrained (so to speak) in the business plans of early hemp enterprises—that some advocates felt fighting for flower's inclusion in the 2014

COMBINING KILLED THE RADIO STAR

Farm Bill was going too far. Less than five years later, with worldwide awareness of our endogenous endocannabinoid systems rising and network health gurus like Sanjay Gupta declaring cannabis safe, the flower market overwhelmingly dominates the industry.

The situation is suddenly so lopsided in favor of this one part of the hemp plant that in 2018 state hemp regulators in New York closed hemp applications to all but those who wanted to grow for any application *but* CBD. Now, it's almost never a good sign when a government agency dictates to farmers what to grow, but I respect New York's effort to say, "Hey, gold rushers, how about a long-term play in, say, hempseed, a genuine superfood that heals soil? Or maybe show the fiber a little love."

The lesson here is not "flower rules all." Rather, it's "just wait another five years." Flower (or at least CBD) dominance, too, is temporary. Get ready for many such major shifts in coming decades, resulting in hot markets that we can hardly imagine today: lignin in hempseed hulls (today almost valueless), hemp plastics from fiber, and some astounding properties in an as-yet-unnoticed cannabinoid #42. (Maybe it will make hair grow; perhaps it will be an undeniable soil cleaner or water purifier.)

How does this affect you, if you're a nascent entrepreneur? For one thing, you might want to avoid ordering too many labels for your initial product—you're likely to make some ingredient changes before too long.

But that's just the tip of the iceberg. What if you invest $25,000 in a cold ethanol processing machine, only to find that the obvious best way to extract whole plant cannabinoids and terpenes at 20 times the efficiency, which won't be invented until 2023, costs $500? This hypothetical (but not far-fetched) dilemma is part of a wider reality in the hemp industry: Our hemp-processing infrastructure and market delivery pipelines are not yet mature. To say the least.

Oregon's Rick Kiyak-Boughton, 59, makes a hemp-and-mushroom-infused chocolate called Green Goddess. He's one of the survivors who was on board the hemp train back in the 1990s. This is how he describes the main cause of hemp's inability to launch to the next level back then: "Folks would go to Thailand or Hungary and come back with fiber swaths that they'd make into fantastic bags and shirts. But when a store said, 'These are great, we'll take ten thousand,' well, they'd already taken their trip abroad. When they tried to reorder, there was no consistent supply chain or pricing and almost no domestic processing."

This inability to pivot (in this case, scale up) still accounts for the principal logjam of the young industry. It explains why so many folks you see working in hemp appear to have aged so much between year one and three, like a second-term president. There are high entry costs for fiber production in particular. To this day, I don't think I've been able to buy a favorite pair of hemp slacks twice, with the exception being hemp yoga pants made by Rawganique of Blaine, Washington.

But what seems like a major challenge can be a great opportunity. Thanks to the democratizing technologies provided by the supercomputers we all carry in our pockets, sides of the business like distribution, accounting, and labeling are a lot easier for an independent to tackle today than they have ever been. These are tools that Kiyak-Boughton's colleagues didn't have back in the '90s.

Still, executing plans in the real world is a lot different from daydreaming about them. Just to get my product into the local food co-op and a couple of spas has involved a ton of time, energy, and bookkeeping. All told, your smartest move, as my river guide trainer advised, is to keep an eye on three turns downstream. Plan for some kind of plan B to the extent possible.

Plan B is pretty much inevitable. It's not so much that you shouldn't put all your eggs in one basket. It's more that you could one day soon have to migrate to a whole other basket, leaving what you thought were the golden eggs behind. You might even want to have a plan B column in your expense spreadsheet.

This initial amusement park ride of volatility and adaptation is typical in any promising young industry, and shows how diverse and valuable hemp is. Take Silicon Valley in the 1970s. For every Apple, Oracle, and Google, a thousand other enterprises, with ideas good and less so, went away for as many reasons. With a bit of brainpower, you'll always be able to figure out something that is first, better, or different, if you're paying attention.

Equally important is to avoid the kitchen sink marketing plan. I get worried when I see colleagues offering 96 products at their trade show booths: soaps, oils, tinctures, snacks, dog treats, bedding. Pick one. Maybe two. You can always expand.

Plan B is on the horizon in my own enterprise, now that I plan to make my initially topical Hemp in Hemp product an edible one. When that happens (cue baroque organ riff), I'll have FDA food regs to deal with. Say what you will about the DEA being out of the hemp game (I say it's fantastic), many of us are waking up to a brave new world known as the big leagues.

Plan B thinking has a long, proud history in American hemp. It goes all the way back to George Washington's business plan. In 2018, while I was wearing colonial garb and wielding that sickle, Mount Vernon estate horticulturalist Dean Norton opened my eyes about the first president's entrepreneurial MO.

Dean is a fountain of knowledge about George Washington, from how often he and Martha bathed, to his favorite mode for making whiskey. As we strolled the banks of the Potomac during a break in the harvest, he told me something that made me think the father of the nation had time-traveled and consulted with Dolly Parton.

"President Washington's most profitable private business was fishing," Dean said as the team of robust Mount Vernon oxen plowed nearby. "He pulled millions of pounds of sturgeon every year, right from this spot. And guess what his nets were made from?"

I was able to guess. But I was still shocked. So shocked that I nicked my finger with my sickle, thus donating nitrogen to another farm's soil. "You mean he wasn't growing for food or clothing?"

"He did grow for textiles, for a time," Dean said. "But his most lucrative use of the harvest was for fishing nets, so he switched to that. The nets were woven from hemp that he grew right over there, by the oxen."

Colonial pivot, baby. I find this consoling.

CHAPTER TEN

How Would the Shaman Bottle Hemp?

Vermont, 2019

One of the best pieces of advice I ever got was from a horse master. He told me to go slow to go fast. . . . We live as though there aren't enough hours in the day but if we do each thing calmly and carefully we will get it done quicker and with much less stress.

—Viggo Mortensen

"What's new on the last-minute complications front?" I asked Colin Nohl by way of hello, tapping snow off my boots and hat. I did not do this gracefully, but I did do it gratefully: I had finally made it to the Giguere family sugarhouse, in the midst of a March 2019 blizzard.

When would I learn to rent the right kind of vehicle in rural America? I had been ice-skating for hours in another subpar crossover since my flight got rerouted from Manchester, New Hampshire, to Albany, New York, in the middle of the night. No big deal, just 7 hours late and 150 miles south of my Northfield, Vermont, destination in whiteout conditions. We had a commercial kitchen reserved for the morning, our arrival now pushed back half a day. Ever the optimist, I still stopped by

Cary's barn, four miles from the sugarhouse, to grab a big vat of hemp flower before catching up with my pards.

A Spoonful of Maple Syrup Helps the Processing Hassles Go Down

Now that I was safe at last in the Giguere sugarhouse, the maple smell and stove heat transformed me from frozen fingers to shedding layers in 10 seconds. My eyes focused. The entire clan was inside, Erin shoveling wood into the boiler, Cary and 77-year-old patriarch Conrad precariously nudging the chimney into better position from a sideways perch eight feet above a hundred-gallon vat of sap. Cary's wife, Kristin, and some cousins were scattered around the boiler, looking nervously up at them and offering periodic suggestions.

I inhaled contentedly. This was all very promising. If you've ever been around maple syrup at its moment of birth, you'll understand why I say that the white-knuckled drive was totally worth it. When an enormous batch of syrup is ready, someone opens the sap pan tap and fills a thermosful. It's served hot in Dixie cups, or, in the case of direct *Mayflower* descendants, shot glasses.

Maybe it was knowledge of syrup to come, maybe the strict hemp-heart-and-cacao diet I had been following during the interstate drive was helping maintain equilibrium, but I was relaxed about the 2 hemp-processing days to come. I knew that few elements of our long preparation for these next 48 hours would prove to be as dialed in as we had planned. Heck, this time plan B had begun before I even landed. I mean, how often does your airline announce a new destination state during a layover?

But if enough did go as we hoped, these days had a dual purpose: (1) preparing 1,000 bottles of Hemp in Hemp and (2) jarring up a cauldron of Colin and Erin's Vermont Farmacy Herbal Salve. Both products would contain seed oil from our joint organic harvest.[1]

Sure, we were already supposed to have the harvest half bottled. But I never doubted our run would get done. I was ready for whatever the fun-loving universe threw at us along the way. The steeplechase resumed before I got out of my wet socks.

"I wish I had better news," Colin said as I sat and started to disrobe. "But the folks over at Victory say our seed-oil pressing has an unusual taste."

I heaved an immense sigh of relief. I had experienced this before. Victory Hemp Foods were the folks pressing our seed oil for us—they had a facility in Middlebury, Vermont.[2]

"That's not bad news," I said. "First, better, or different,' baby. That's Samurai. It's not unusual to my taste buds. It's just more full-bodied than the seed cultivars most North Americans have been pressing so far."[3]

I stood up to step out of a layer of long johns and pointed to the sap pan. "Like amber versus very dark syrup. Stout versus pale ale. A hemp flavor for every palate."

"Okay. Cool," Colin said, nodding. Then his smile faded a notch as he noticed the pronounced limp I was sporting. "Did you remember what I told you about the barn floor?"

Needless to say, there had already been additional complications during those first hours in Vermont. It was only as I skidded into what I hoped was the side of the Giguere barn 2 hours earlier that I recognized the magnitude of the shoveling that would be required just to reach our harvest. I was pretty sure that I could tell where the barn door was. While I dug around in my luggage to find my mittens, I was relieved to see the Yeti-like form of Cary approaching, carrying two shovels.

"Watch out, that floor is a hockey rink—really, really slippery," the 31-year-old Colin had warned me on the phone when I'd pulled over for a stretch outside of Killington on the drive north. "You won't see the ice, because it's so clear and flat."

Then, just a few hours later, the moment we'd chunked out the barn door enough to pry it slightly open, Cary repeated the warning. Right down to the two *really*s before *slippery*.

And still—unsurprisingly, to anyone who has spent an hour working around my Clouseau-like life—on my second step I slipped magnificently and found myself resting on what I know from my kids' Spanish lessons are called *las nalgas*. Actually, only on the left nalga. The right one was wedged against one of our Samurai flower-bin shelves, millimeters from knocking 150 pounds of it onto the icy floor. A good portion of 10 months of work by four people and tens of thousands of plants came inches from spilling.

When I related the anecdote to Colin in the sugarhouse, Cary summed up hemp entrepreneurialism and possibly life by shouting from his ceiling roost: "I learn every lesson the hard way. Including the slippery-barn-floor one."

Just as golf is really three games (driving, short game, and putting), so you are essentially starting spring training afresh on the first day of any new stage in the hemp process. Getting the cobwebs out. That's how I looked at that resulting bruise over the coming days: as a reminder that we're all beginners, every time.

———————

Fresh maple syrup helps you forget a bruised body. In Vermont, sugaring is the key family and community bonding time. I think the idea is, *Cabin fever is setting in, might as well all gather together so we can have witnesses. Plus, let's see if we can't tone it down a notch by flooding our systems with glucose.* I think, overall, it's a good system. Any vestige of ritual keeps a culture alive.

"It's the sugaring and the autumn wreath making that keep this family together," Kristin, who is an early preschool educator when not sugaring or making wreaths, told me.

This is the same "when it's time, it's time" role I hope my ranch's goat rituals and hemp cultivation inspire in my progeny. And rural Vermonters are learning fast that sugaring can overlap with hemp processing. For my part, the processing window was narrow. I had a series of live events coming up in the ensuing weeks, plus a waiting order for Hemp in Hemp from my New Mexico food co-op. And I was out. Even at home we were down to our last bottle. Erin and Colin were getting low on their salve, too, which (on-brand alert) infuses its cannabinoids for a full moon cycle.

For the 2019 Hemp in Hemp run, I was scaling up slightly, but once again processing what felt like a manageable amount. This batch of three-ounce bottles would, when it sold through over the course of probably a year (given my limited time for marketing), gross my family $50,000. I'd estimate the eventual net as $20,000. We give a lot of bottles away. Then there's the cost of bottles, caps, graphic design, labels themselves, web work, testing, permits, organic certification, accounting, legal, storage, and shipping (but not genetics, because *we own them*, as free farmers always have). And now I'd be sharing revenue with my new retail partners out West. Part of the five-year plan.

The question I wrestled with in the sugarhouse was, was I really about to extract two family members from their clan's principal annual activity, for 2 entire days? There was a septuagenarian swinging from the ceiling, for heaven's sake. But everyone at Château Giguere seemed to understand a key point: that it's wise to feed both the sweet tooth and the endocannabinoid system. These are two essential dietary services for humankind.

Extraction Methods Old and New

The best part of processing hemp for 2 days in the mountainside commercial kitchen that Colin had booked was seeing and smelling two hemp product lines, derived from the same crop, bubbling forth in side-by-side cauldrons (okay, monster pasta pots). I'd even call it a top 20 moment in my working life.

I love the crazy shift in scale that occurs when you go from being a customer to a producer of anything—broccoli, hemp massage oil, compostable hemp-plastic paper clips, whatever. Instantly it's about tons of product, rather than "Honey, did you bring the produce bags?" A hemp customer for two decades, suddenly I was in an apron and gloves with 1,000 empty bottles smiling up at me from a shiny metal counter. Even with my family's considerable omega-3 and cannabinoid diet, I thought as we unboxed that sea of glass, this was an amount I would be hard-pressed to argue was for personal consumption.

No longer would we be buying 16-ounce bottles of Nutiva hempseed oil, grateful as I've been for this superfood for decades. Now we're thinking in terms of rodent-proof storage bins for multiple tons of hempseed and flower. When I first trudged up the un-shoveled kitchen path on the far side of Northfield, I was just excited to make some more of the product that I most wanted to use. When I carried full bottles down the same path 48 hours later, with the roof eave icicles a few inches closer to the ground, I was a craft commercial provider.

Another reason this hemp-processing run ranks so high is because the product that wound up in those bottles represented the regenerative economy manifesting in the real-world marketplace. I had the opportunity to practice, in the processing stage of the season, everything I'd learned about plant pace over the course of the year. It was my hope that the resulting mind-set, method, and love would show up in the product.

It's no accident that we speak of returning to our roots. The moment I accepted that I was about to dedicate 2 days of my life to preparing product, I relaxed. I was shifting out of fifth gear even as we fired up the mammoth stovetops. Confront a human living at multitasking, finger-swiping speed with the methodical pace of a plant's life cycle, and he often finds it pleasant. I think this is why so many military veterans from George Washington himself to today's resource war vets rediscover their sanity in farming.

It's also a quality-of-workmanship issue, taking pride in your craft. The way I like to make hemp products is deliberate, time-consuming, and small batch. This is not to discount the many positives of digital-age conveniences. The key now is, like good primates, to learn and build from the obvious missteps of the industrial centuries, including the main one: not embedding regenerative processes in our society's economic and agricultural pipelines.

Accordingly, my first question when embarking on any stage of the hemp process is, "What has always worked?" Another way of asking this is, "What would the shaman do?" And what humans have always been able to do in order to extract the healthy, tasty, or otherwise useful parts of a plant is to heat it up in a lipid.

Thus, when attempting to capture these values in a bottle, as I've already revealed, I'm an unapologetic fan of decarboxylation. I'll relate our step-by-step process on that mountain in Vermont, which basically involved combining the hempseed oil and the flower from the same plant, heating the concoction, and subsequently becoming Druids. But to provide options, I'll also run down the other four of my personal top five regenerative cannabinoid/terpene-flower extraction methods.

The three biggest factors that will affect your enterprise's choices are the quantity of product you'd like to produce, how quickly you'd like to produce it, and equipment costs.

DECARBOXYLATION

When you heat hemp flower just right, you remove a carbon atom from your cannabinoid molecules—such as CBD, or the one I so love seeing come back in testing samples, CBC—hence the term *decarboxylation*. You'll recall that this atom removal transforms a cannabinoid from its acid form to its active state.[4]

The procedure is so tried and true, so ancient (even being alluded to in biblical priestly anointing-oil passages), that I wonder if the term *processing*, given its association with slices of prepared cheeselike product, is apt. It just doesn't feel true to the "double, double toil and trouble" life that we lived for those shivery 48 hours in that 2019 storm.

The other component of decarboxylation that I love is that you transfer all your cannabinoids and terpenes in pretty much the exact levels and ratios in which they are found in your flower. Given that in five years (tops) some piece of well-publicized research is going to launch that cannabinoid #42 we'd never heard of to Next Big Thing status, I'm a devoted fan of this mode. I believe it's likely that many of us benefit most from the interaction of these hundreds of compounds in a "whole-plant extract."

By contrast, as we touched on earlier, some early CBD customers are being trained to ask, "How many milligrams am I receiving of this one cannabinoid per serving?" That might be relevant on the explicitly medical side of the plant's applications. Soon, though, as we begin to connect the dots on the nutritive and health maintenance sides, I think more of us will ask, "What are the cannabinoid and terpene ratios in the source flower, and how were they extracted?"

The bottom-line performance I experience in my own body and mind is the main reason why, in a volatile cannabinoid market, I utilize the whole plant. And it's the reason I don't discuss cannabinoid isolate much, other than to disclose that current gold rush prices for CBD isolate are hovering around $20,000 per kilo. Under ideal conditions and with extremely efficient processing, you could produce 20 or more kilos on an acre of flower-only hemp if your cultivar produces 10 percent CBD.

Remember that if you don't own the processing equipment, you'll either pay for toll processing or have to sacrifice a percentage of the harvest, just as during feudalism. Still, we're talking about a six-figure gross per acre, if current prices hold. Some folks, as we've seen, are already in the market for emphasizing or isolating cannabinoids other than CBD, such as CBN. For his part, Edgar is keen on CBG, because chemically, it's the building block for all other cannabinoids. The source code.

"All the other cannabinoids regurgitate from that one," he told me way back in 2016. "So if we breed for that, we stay in the game no matter what other cannabinoids become desirable in the marketplace."

As CBG is already worth $25,000 per kilo in today's market, Edgar was not swimming blindly. Because fewer farmers are growing for it, we might see its wholesale prices hold longer than CBD's inevitable correction as markets mature and supplies increase.

Notice that Edgar speaks of "breeding for" a particular cannabinoid profile. Not isolating it via machinery. He enjoys, as do I, working toward desirable components as part of a cultivar's whole-plant profile. I'd love to develop a cultivar whose flowers display CBC in a one-to-one ratio with CBD and some of the funner terpenes, like myrcene and pinene. It's just my preference—I love grape and pine tree scents—a breeding mission onto which I stumbled by chance (or plant's choice) in my own hemp diet. Everyone's body is different and no one's body chemistry is static.

Okay, there you go. I've disclosed that a market exists for isolated components of the cannabis flower. And that in most circumstances, I wouldn't spend my money on these isolates. But then I am in the entourage effect business, not the CBD business.

COLD ETHANOL

One reason that cold ethanol extraction is on my approved list is that it's still a pretty tried and true technology—indeed one of several processing modes the FDA considers "low risk" when done right—but it's a faster, larger-scale mode of getting those cannabinoids from the cannabis flower.[5]

The end result of running raw flower through the ethanol process is crude, that luscious, sticky caramel goop we discussed earlier. This is isolate's precursor. Like decarboxylated flower, crude contains all your flower's original cannabinoids, though much concentrated. It can go into your own products, be sold to others who want to isolate the CBD with chromatography equipment, or be "white-labeled" to folks who will add your crude as an ingredient in their own products.

I believe products resulting from ethanol crude processing can be bioavailable, and I think the process is doable in accordance with regenerative practices and values, especially if your facility is solar-powered. Energy choices are a vital part of the process for the regenerative entrepreneur.

A cold ethanol processor can be a tabletop unit or, as with Dexter Rice's ceiling-high behemoth at Nature's Love, much fancier. Either way, modern ethanol processing is when you take off the shaman's tunic and broach the alchemy era. You start busting out the beakers and plugging things in. You check pressure gauges and deal with flammable materials.

Margaret Flewellen uses cold ethanol extraction for her Natural Good Medicines line of tinctures in Oregon. I've studied and filmed her processing MO since 2015, starting when she had the simplest model from Eden Labs in Seattle. Here's how this kind of processing works: Margaret loads her ground-up flower material (grown 100 yards away by Edgar) into a finely netted bag. The bag goes inside a glass dome containing a condenser. Very chilly ethyl alcohol (–20°F) rains over the flower. The ethanol (C_2H_5OH) acts as a solvent, removing the cannabinoids and terps that reside in the flower's trichomes.

Upon filtration, the process concentrates these to about 50 times the level of the flower itself, but in the same ratios. The alcohol filters out via a vacuum process, and can be reused multiple times. There are some other nuances involved in the process, such as whether you want all your plant's chlorophyll and waxes removed from the crude. If you do, the extracted product is often called *full-spectrum* rather than *whole-plant*.

One option a bootstrapping enterprise might consider is devoting some of its flower harvest to producing a value-added product and some to producing wholesale crude. This is a viable hedge if wholesale prices hold for a few more seasons. I know several farmers who employ this type of strategy. Plus small-scale cold ethanol processing equipment can be feasible to own for a self-funding entrepreneur.

For Margaret, the hardest part of ethanol processing is managing growth. Natural Good Medicines has upgraded processing equipment three times in five years as Margaret's business has expanded.

"It takes about eight hours to process forty-eight pounds of flower in our latest machine," Margaret told me. "We lose about ninety percent of the mass in the extraction process, leaving us with a crude that, depending on the source flower, is roughly seven percent THC and seventy percent CBD, if we start with flower that has .2 percent THC and eleven percent CBD."

Embedded in these numbers is the reminder that the ethanol process concentrates all your plant's cannabinoids, including THC. Assuming you don't want higher THC levels in your product (meaning you aim for your product to meet current hemp definitions, something ready for customers of all ages), you have to dilute it back down.

Margaret said it's not hugely expensive to purchase a starter ethanol unit ($11,000), and the prices can get up to $275,000 for the largest commercial-scale unit. Under the guidance of engineer Bill Althouse, the folks at the Fat Pig Society have jury-rigged their own ethanol processing technology. Dexter Rice's $175,000 unit processes 175 pounds per day.

"Operating according to GMP [good manufacturing practice] standards for food-grade products is a cost and a challenge I would warn anyone considering a small processing operation about," he told me. "It's not that easy to self-learn—I wouldn't underestimate the time and energy that go into it."

This sage wisdom is true for all processing modes. In the final analysis, the big advantages of cold ethanol processing are volume and speed; you can probably make 10,000 units in the amount of time it takes me to make 1,000 of my decarboxylated product.

WATER OR ICE EXTRACTION

This type of "solventless" extraction has a place. Ganja processors will know it as the bubble hash process. What it does is remove the trichomes from cannabis flower with water and elbow grease. The flower is physically stirred (by hand or machine), and the water or ice is filtered out via screens or mesh bags.[6]

COPPER STEAM DISTILLATION

Distillation passes the shaman test and can provide a nifty mode for extracting alternative components of the flower, such as essential oils. I feel so bad for my colleague in Hawaii, Lelle Vie, who is such a mistress at this mode that folks always want demos, so she's forever toting her giant copper distiller to conferences. More than once I've crossed paths with her at an airport or farm-side event and found myself carrying shards of copper sheeting for her.

PRESSURE (ROSIN) EXTRACTION

Lastly, if you really want to be an opposite-of-isolate processor, the pressure extraction mode, used for centuries by the kief-makers in the hills of Morocco's Rif region, can hardly be topped. You apply just the right amount of mild heat and tough love (with a press or, in some traditional modes, by beating wrapped flower with a stick) to extract the desirable oils from cannabis flowers. Remember our farmer friend we met at the United Nations, Adebibe Abdellatif? That's his production mode.

Here in the West, the resulting product is known as *rosin*, and it's increasing in popularity as a high-end, solvent-free processing mode. There are easily a half dozen commercial rosin presses on the market. Though you might want to head to the Rif region to see how it's always been done.

A processing mode that doesn't fit as well under the whole-plant umbrella is CO_2 (sometimes called *supercritical*) extraction. That's because while it's very high volume and a clean method when it comes to solvents, the process can remove too much of the plant's terpene profile.[7] You can re-add terps and other desirable parts of the plant later, so it's not a total disaster. But why use a process that removes 'em in the first place?

Off the list entirely are butane, propane, anything ending in *-ane* or *-ene*. Anything done with petrochemicals is a nonstarter. Not just because of potential solvent residue in the resulting product but because we're trying to wind down the dinosaur juice era as a species.

In the end, I employ decarbing exclusively for my own product not just because it's been field-tested for millennia, and not just because you end up with the original levels and ratios of all your cannabinoids, but also because as long as you are paying attention and have better balance than Inspector Clouseau, it's so dang simple. It's just heat. How much heat and for how long matters a lot, though. Also you produce far less product than most other modes, and it takes up a lot of time. So decarbing is overall best suited for small-batch, top-shelf products. Which is what I make commercially and personally. It's what I eat in the morning goat yogurt. And it's what I rub on muscles sore from corralling infuriating goats and much-less-maddening hemp work.

Now we're ready to reveal how we removed that carbon atom to create the Hemp in Hemp product run in 2019.

Goop Is Good

Among the supplies Colin, Erin, and I toted into our kitchen were 27 gallons of hempseed oil, in six 4.5-gallon "jibs," fresh from the Victory Hemp Foods presses.[8] Hemp in Hemp being both a seed and a flower product, we had a multistep process in front of us prior to bottling. First prepare the seed oil, then infuse the flower in it.

Experimenting with hempseed oil as a decarb lipid was pure Dolly Parton: The result of having access to a farm-side seed-oil press and not having many other options. Organic coconut oil would have been very expensive. Otherwise I might not have discovered what an ideal (and low-carbon-mile) base hempseed oil makes for a cannabinoid/terpene product. Because we're working toward an edible product in future seasons, what's really exciting is that the combination of seed and flower encourages both superfood and health maintenance qualities. Plus, how fun to utilize more than one part of anything these days, let alone a plant.

When coming up with the idea for Hemp in Hemp two years earlier, I didn't know anyone else who was immersing flower in the seed oil from the same crop. Now it seems to be catching on. Our friends at Salt Creek use hempseed oil, albeit not from plants they grow, as the base for their whole-plant 42 brand capsules and Eleven Acres products (they grow the flower). I hope many more join the parade. This is the open-source era. Feel free to re-create Hemp in Hemp in your ecosystem, in your regional economy, with your cultivars. If you grow organically, I'd love to try your offering.

Following the snowy 2018 harvest, as with previous seasons, we first had to press our seed. Actually, first we had to send off a sample to make sure it passed its microbial test, which it did. As we no longer had access to a seed-oil press, I got on the horn with Chad at Victory Hemp Foods.

As with ethanol processing of flower, you lose the majority of your mass in the seed-pressing process. So you get, say, 40 gallons hempseed oil from 400 pounds of seed. Our seed oil was darker than what you find in bottles of most North American hempseed oil options at the moment. Holding a clear bottle of it to sunlight, what I see reflecting back is luscious, olive goodness: all 20 amino acids and three essential fatty acids.

Dark though it was, our 2018 oil was not as dark as the earlier pressing of Hemp in Hemp. That's because in 2018, the seed we used wasn't exclusively our Samurai. Since the seed itself was both limited and in demand for its genetics, we asked the good folks at Victory to mix our Samurai seed with their US-grown organic hempseeds for the small-batch pressing.[9]

Still, the decarbed product had that chartreuse hue. This special darkness helps explain why Chad's plant manager wanted to alert us about what to his peeps was the unusual resulting taste profile. Samurai is an unusual cultivar. *Robusto* is the word that comes to mind. Sustaining.

The final Hemp in Hemp mixture was also delightfully cloudy. That's because I asked the folks at Victory to leave our hempseed oil unfiltered after pressing. Another way I like to sidestep the gold rush herd is to leave the plant's lignin in. I believe it has beneficial qualities. More than one industry professional advised me against this, saying it reduces the product's shelf life. Even if true, I'm fine with that. In fact, I hype it. On the label, I recommend storing Hemp in Hemp in a cool place and using within 90 days. I even added a line reading SHAKE WELL: THE GOOP (LIGNIN) IS GOOD.

These are just further reasons why it's so vital that you own and develop your own genetics. Just as every fine vintner cultivates her own distinct grapes, your taste will come only from your hemp. This concept of "terroir" is essential to the top-shelf hemp craft model.

Now we had our lipid base and what's more, all of it made its way up the slippery path to our kitchen without spilling. (The charming, remote spot was at a resort called the Woods Lodge that used to be a girls' camp called Camp Wihakowi.) After spending an hour making sure we and all our surfaces and cookware were clean and up to code, we were ready to heat 20 pounds of flower in those jumbo pasta pots full of seed oil. Thus begins decarboxylation.

I had brought 50 pounds of flower from the barn, incidentally, once I had got off my nalgas and learned how to skate across its sheer floor (long, relaxed-as-possible strokes). That was just in case we got carried away and wanted to bottle a surplus for a future run. Or, more likely whenever I am in a kitchen, because stuff boiled over and we needed to repeat a step.

For the kids reading this who didn't come of age under prohibition, can I just say what a pleasure it is to drive around in a rental car with out-of-state plates toting 50 pounds of fragrant cannabis flower with nary a worry about law enforcement attention? Cannabis legalization overall is a simply fantastic development for the American economy and for public safety. And Vermont's embrace of it—at all THC levels—has provided peace of mind on numerous occasions, including during a chat earlier that first processing day with a state trooper. He wanted to let me know that my rental car's running lights weren't on—he couldn't have cared less about the bin of terpenes and trichomes in the backseat, on my sweater, and on the dashboard.

I thought, as I pulled on my apron, that Vermont is the West-est state East of the Mississippi. Just remember, deciders in the South Dakotas, New Jerseys, and Mississippis of the world—your residents have endo-cannabinoid systems too.

Double, Double Toil and Entourage Effect

Decarboxylation aficionados all have their favorite modes. The process is a rarity in modern life: both a science and an art. Colin and Erin usually heat the flower for their products in an oven before getting shamanistic in the cauldron, in order to reduce processing time.

Myself, I'm as old-school as it gets: For the 2019 batch I filled two giant pots with hempseed oil, added pounds of flower to each, slowly raised the temperature, and decarbed at a steady 220°F for 3 hours. This reflected a lowering of temperature and increasing of processing time from the 90 minutes at 230°F I had been running. I did that because hempseed oil starts losing some of its nutritive properties at 190°F and I didn't want to cook out the goodness. A good portion of each three-ounce bottle would contain pure hempseed oil heated to 185°F, so I wasn't too worried; plus, as of 2019, the product wasn't yet intended as food. But in discussing decarb efficacy with colleagues, I thought longer heating at a lower temperature was worth exploring.

I'm a happy man when I am stirring a cauldron, breathing in the essence of cannabinoids for 180 minutes. I think I might have some Druidic lineage.

"Watch me try to stop smiling for ten seconds," I said to the 30-year-old Erin, who was stirring one of her salve pots with a giant ladle next to me.

"That's the terps," she said, waving a couple of be-gloved fingers in front of my face. "Your body's getting ready for you to introduce cannabinoids. Receptors are springing up."

One of the coolest parts of the decarb process for me is watching the carbon molecules start bubbling off the flower-and-seed-oil mixture. These are smaller, more numerous bubbles than you see in actual boiling, more like the tiny ones that bubble off your carbonated beverage on a summer day. I'm extremely grateful to Bill Althouse for teaching me to look for this distinct bubbling pattern way back when he was still tired from my saxophone wake-up call, because it's a great way of ensuring that your concoction doesn't get too hot.

That's important because when you get beyond decarb temps and into the boiling range, all your cannabinoids can get transformed into CBN. This isn't necessarily bad, if a CBN-enhanced product is your goal (as it is in some of Erin and Colin's products). CBN is often used for pain reduction or as a sleep aid.[10]

But since I was endeavoring to highlight all of Samurai's cannabinoids and terpenes, the reality was that there in the hilltop commercial kitchen, with yet more snow socking us in, one of us had to keep stirring and watching our thermometers the whole time. Vigilantly. It can be a fine line once you're approaching that 225–230°F degree range. Terpenes, in particular, can be fragile at these temperatures. If one of us wasn't on stirring duty, she was dashing around, preparing bottles, working on the salve product, or cleaning up after my latest spill.

Before I knew it, we were done. When the timer went off and the mixture felt and tasted ready, we punched out. Bright and early the next morning, all of us still in a terped-out state of mind, Colin taught me a great time-saving bottling tip. Though a relative youngster, he has a decade of food service management experience under his belt. He explained that if we "hot-bottled," meaning jarred our product above 185°F, we'd be pasteurizing as we went. This saved hours on the thousand-unit run, because in previous years I'd washed, dried, and heated all the small maple syrup bottles before filling them.

Like every part of the hemp season, like every part of life, processing has a lot of karma embedded in it. I am still bubbling over with gratitude now, several months later, as I relate that every step of this batch of Hemp in Hemp bottling went just right. The decarbing part came off without a hitch, and on bottling day, I spilled only a few gallons of product.[11] (The low spill rate was largely thanks to the handy, hobbit-sized funnels we used.) Erin and Colin's moon cycle salve also went smooth as butter. Literally. Try some.

While we were toting those 100 boxes full of bottled cannabinoid/ superfood offering through the snow, all of it certified organic, I was beaming. I was excited about getting the stuff out to the world. But first and foremost, the Fines back home on the Funky Butte Ranch, the people who are irreplaceable to me, were all set in the hemp category of our superfood health maintenance regimen for a year. Another primeval box checked off for this aspiring neo-Rugged Individualist. From a crop I grew.

After loading the rental car to the sprockets and skidding down the mountain for the last time, I shipped off a couple of boxes of both my product and Colin and Erin's to Colorado, where they would be waiting for me at my speaking event and book signing a week later. And I sent a few bottles off for cannabinoid testing. They passed.

Colin and Erin snarfed some of the ceiling-high stack of pizzas I'd brought in appreciation for their generosity of time, stirring muscles, and knowledge, and then dashed back to the sugarhouse. Cary and I reconnoitered and drove a couple of hours north to the Orwell, Vermont, library to give a talk to interested farmers near the Canadian border.

Back in Northfield much later that second sleepless evening—a gorgeous one, the icy sky having cleared to reveal a glitter box of stars—we all converged at Château Giguere. Cary opened the bottle of still-warm Hemp in Hemp I had given him and poured generous portions in brandy glasses for the whole family.

"To USDA-certified organic craft cannabis," was my toast.

The 2018 vintage tasted scrumptious, even better than the first batch, was the uniform consensus. Cary's observation was that the nutty finish reminded him of the polyphenols one tastes in high-end olive oils.

It's possible the judges were biased: This is a stout family when it comes to beer. A "Grade A Very Dark" maple syrup bunch. Also we had all grown the crop we were enjoying together. I didn't overthink it. I finished my shot of Hemp in Hemp and then got started on the maple syrup.

———————

I wound up returning my rental car in New Hampshire, which had only 76 miles on it when I picked it up in Albany, in a very different condition from its original one. The fellow at the airport return department seemed unsurprised and unperturbed. Maybe they had hemp farmers coming though here all the time. Regardless, I was grateful that the smell and back road grime didn't raise eyebrows. Even the travel nightmare was going as well as could be expected.

As I checked in for my return to the West, I thought some more about scaling up. In 2019, three people's lives were dominated by this micro-processing run for 2 full days. And in any regulated workplace, our punched-in hours would really have added up to 3 days. Or 2 with a ton of overtime. Not counting the 10 months of work that preceded processing. All for 1,000 units.

I couldn't help wondering what the whole hemp cycle would look like now that I would be working with colleagues with a built-in retail outlet. Based in Arizona, Kim Williams and Dana Rae Zygmunt were my new Hemp in Hemp partners. These two women were proven righteous humans whom I've known for years. They came from medical backgrounds, and, some would say most relevantly, they had an existing shop for our product line in the thriving Tucson health center called Tumbleweeds they founded and had operated for seven years.

Around the time I was processing in Vermont in 2019, Kim dropped casually that she believed interest in Hemp in Hemp at Tumbleweeds would reside "in the two-hundred-bottles-a-month range, conservatively."

Yowzer. That would mean monthly sales at 20 percent of current annual sales. At one location. Soon I was really going to have to turn pro. I'd already dealt with hemp permit paperwork, organic certification, and, ya know, growing fine crops. But food-grade certifications and distribution chain of custody rules hovered on the horizon. I hoped then and still hope we can avoid the homogenization that so often erodes what's awesome about craft products when they scale up and need to meet different regulatory guidelines. Would late night shamaning-out in icy mountainside kitchens still be in my production SOP? If it were up to me, even if we had to do what we did in that Vermont kitchen monthly, I'd still prefer to process by hand in a cauldron.

Could I, though? With each expansion there is more to think about on the business side. Take batching, for instance. Here we have another process about which Colin and Erin had been teaching me between sing-alongs and product stirring. Every bottle's label has to be tied to its ingredients' source. For me, to date, it's been easy: Batch 1, 2019, that's it. One crop, one product run per harvest. It's the blessing of a simple recipe produced in a single batch.

But Colin and Erin, who process a dozen times per year, have to keep track of every bottle, its source flower, its source seed oil, its source eucalyptus and lanolin (and whatever else Erin adds to the moon cycle salve). My colleagues do that with a UPC symbol that accompanies their labels. They even ward off scrutiny with one of those "This product has not been evaluated by the FDA and we make no health claims" disclaimers.

Since I have only one retail outlet among the three places Hemp in Hemp has so far been marketed (my New Mexico food co-op), I can for now let those good folks attach their UPC label. I could say "our" UPC label, since I'm a co-op member. I met the mother of my children next to the produce aisle. The other places one can find Hemp in Hemp in these early days are at a spa on Maui and at my speaking events. That will change by 2021. At least that's plan A.

Batch management and claim avoidance are just the start. I'm equally concerned about maintaining control of ingredients. I once did a radio news story about a local who was taking her regionally popular salsa to the big time. Her one complaint was, "I'm forced to use facilities that only accept certain suppliers for my core ingredients."

All of which is to say, once again, be careful what you wish for, hemp entrepreneurs: Between GMP quality control, FDA regulations, and payroll paperwork, even with your product ready to ship, you might find your dream job quickly becomes a real job.

Still worth it, I thought, watching my club soda decarboxylate from my plane seat. If designing labels and planning for inspections were not exactly what I thought I'd be doing with my life when cannabis became legal, neither was producing a fine omega 9-6-3 balanced product.

When it came to my own entrepreneurial effort, I just hoped that no matter when and how we took it to the next level, I'd still be in the field more often than not. More immediately, I was finally going to be able to call it a day. I thought about this with relish as the final flight banked toward Albuquerque (I'd detoured to Colorado for some research into hemp cooperatives after leaving Vermont). I'd share trampoline-jumping energy with the kids in exchange for their granting me some precious grown-up time with my sweetheart. Then I'd milk the goats and toss some Hemp in Hemp into the bath. I knew from experience that I'd barely hang my towel before I was ready for the dreamless. Health maintenance is so much better than any kind of recovery.

CHAPTER ELEVEN

Hemp Keeps You Thin

Gamma-linolenic acid (GLA) is . . . known to affect vital metabolic roles in humans, ranging from control of inflammation and vascular tone to initiation of contractions during childbirth. GLA has been found to alleviate psoriasis, atopic eczema, and mastalgia, and may also benefit cardiovascular, psychiatric, and immunological disorders . . . As much as 15 percent of the human population may benefit from addition of GLA to their diet . . . It is important to note that hemp is the only current natural food source of GLA, i.e. not requiring the consumption of extracted dietary supplements. Because of the extremely desirable fatty acid constitution of hemp (seed) oil, it is now being marketed as a dietary supplement.

—ERNEST SMALL and DAVID MARCUS, *Hemp:*
   A New Crop with New Uses for North America[1]

Snacks? What kind of snacks? Are they good snacks?

—HOBBES, *Calvin and Hobbes* by Bill Watterson

ven when I'm on the road, not a day goes by when hempseed is not in and on my body. I pack the liquid Hemp in Hemp in checked luggage for air travel, and I keep hemp hearts and cacao nibs on-call in my carry-on. At home, I eat probably a solid cupful of hempseed in some form most days, and it's in my human and goat kids' bellies by 8:00 a.m.

But my diet is not yet typical. Many of us who are part of the early, self-reinforcing hemp-family bubble tend to think everybody on the planet is a day away from kicking Wheat Thins and switching to hempseed. The reality, as we've discussed, is that 99 percent of North American homes have never popped open a hempseed oil bottle or filled a bulk produce bag with hemp hearts.

A Deserving Diet Craze

So how do we spark the diet craze that will expand our acreage to where our climate needs it? In order to do its part in preventing humanity from tumbling over the climate precipice, hemp has to become a staple crop. That doesn't happen in energy-demanding sinsemilla "grow facilities." It happens on huge numbers of outdoor farms cultivating for food. According to the Environmental Protection Association (EPA), hemp sequesters 14 metric tons of carbon per acre.[2]

How many acres represent a staple crop? Let's start with those 234.7 million acres American farmers currently devote to soy, corn, cotton, and wheat. If organic dioecious hemp acreage reaches that level, we'll be sequestering 3.29 billion metric tons of carbon.

Not to overshoot or anything. We are speaking about a crop currently at 150,000 acres. Our goal might seem to run a bit beyond the 20-acre farms on which we've been focusing. Without a doubt, we are talking about quantity if we're trying to launch a healthy food renaissance. But 14 million farmers each cultivating 20 acres gets us there. That still only gets 4.6 percent of Americans farming. And those numbers come down when you reflect that in states like North Dakota, the average farm size is 1,238 acres. Roger Gussiaas's smallest contract farmer for his seed presses in the Peace Garden State grows 50 acres (a 50,000-pound seed contract, by dry weight), but most are in the hundreds of acres.

Make no mistake: Farmers have the number one job in the regenerative economy. That's because agriculture, which today releases more carbon than all modes of transportation combined, will sequester the most carbon when regenerative modes become the norm.[3] I envision one day flying in a solar-powered, hemp-bodied plane, over section upon section of hemp planted in a checkerboard pattern in rotation with other large-scale food and soil-building crops. All grown organically, all sequestering carbon, unbroken horizon to horizon.

About five years after we hit that target, while you're enjoying the cleaner air and water and stronger rural economy, don't forget to take some pleasure in watching all the pundits currently predicting CBD will be the largest long-term hemp sector quickly change their stories. That's good. That means climate change is being addressed.

To literally seed the supply side, if I held the USDA purse strings (or its equivalent in any nation), I would direct tons of energy, human power, and funding toward a Digital Age Homesteading Act that incentivizes a surge in independent hemp production and other soil-building crops that provide healthy food. The idea fits seamlessly with all this talk of a Green New Deal. It also rebuilds rural communities while all the climate mitigating is under way.

Incentivizing is a start. But the fact is, as I mentioned when relating the impact of my own family's wildfire jolt, we all probably require the single-minded refocusing of economic purpose that last occurred in the United States after Pearl Harbor was attacked in 1941 in order to achieve our goal of immediately transforming our agricultural economy into a regenerative one. Of course, pretty much everyone I meet across the planet who does not hold a public office or work for Monsanto or Syngenta seems to realize this is imperative. But no one wants to experience a collective, climate-based shock. In this case it would probably mean a major city sinking into the ocean.

Even with incentives boosting supply, there must be consistent, enduring markets for all these putative millions of acres of fantastic hemp and companion crops. Billions of people must crunch hemp food every day, instead of today's mere millions.

A couple of years back I had an idea. As an author, I recognize that there are few better ways to guarantee a spot on a best-seller list than to capitalize on a diet fad. The list is always clogged with some secret to weight loss that is linked with a healthy, sexy future. *Celery Juice: The Most*

Powerful Medicine of Our Time Healing Millions Worldwide is dancing with the top 20 on Amazon right now.

This kind of title used to cause me dismay, but I can't deny the genre moves units. As a survivor of the Atkins craze, I have memories of having to fight for my slices of bread at restaurants against significant moral opposition. Remembering the force of that diet fad, I think, a decade later, that it would be cool if we early industry influencers rallied to establish hempseed as the next diet craze, the next acai, the new Paleo. I'm happy to write the diet book.

It helps that hemp will make a legitimate health craze for a lot of reasons, from its ideal omega 9-6-3 ratios to magnesium levels difficult to find in vegan foods. But a lot of foods are healthy. Healthy alone isn't enough for the best-seller lists. The question I've been mulling is: For someone who has never eaten hempseed or hempseed oil, what one beneficial component of its impressive nutritive profile sets hemp apart from the field? What will make millions of people take their first bite? Once they do, I'm confident they will continue biting.[4]

Our friend Chad Rosen, over at Victory Hemp Foods, is an ideal fellow to talk to about this kind of thing. The 40-year-old Kentucky native started pressing hempseed for a 100-acre contract with a farmer in 2015, and today says, "I think I remember when a ton a day felt like a lot." Victory now presses three tons of oil per day.

I called Chad early during my inquiries into this topic of launching hemp food into the big leagues. If his take didn't bring me down to earth slightly, it did educate me to ask the right questions. That's because Chad is not a cheerleader. Rather he is one of the few US folks, alongside Roger Gussiaas and a couple of others, who can speak as a real-world seed-market veteran.

"I burst a lot of people's bubbles when I say that hemp has a big learning curve before it totally remakes the human diet," he admitted to me. "Because we don't really know how folks eat it yet. A spoonful of hemp hearts in yogurt is not a major industry."

"I hear that," I said, spooning a tablespoon of hemp hearts into my yogurt.

He added, "Will there be a dominant taste profile, or many profiles? At the commodity level there are advantages to consistency, although hemp's diversity can of course be an asset at the craft level."

I understood his point: People don't even know what they mean or what they want when they speak about the taste of hempseed. Most don't even know it's technically a nut.

"And here's another difference between hempseed oil and other popular oils. You eat hemp oil raw, you don't cook with it. That's a learning curve for some customers."

3T3-L1 and Your Lipid Neighborhood

Chad's broader message is that hemp needs its "hook." We have truth on our side, sure. We've already got our fellow label readers, our fellow co-op shoppers. You can always find hemp hearts, hemp cereal, and hemp energy bars at Whole Foods. But if we're talking about significantly displacing corn and soy with organic hemp as a human and nonhuman food staple, how do we get there? How does *hemp* come to mean something as pervasive as (and let us pray, even longer-lasting than) *Atkins*? What, I still wondered, would be that best-seller's title?

I discovered the answer in January 2018, on a volcanic rock 3,700 miles west of the Funky Butte Ranch. It happened at one of the tea ceremonies Professor Qing X. Li, PhD (Qing to his friends) was always holding in his offices in the Department of Molecular Biosciences at the University of Hawaii, Manoa. Really good white tea, too; it came loose in beautifully painted tin containers with calligraphic labels.

Qing, biochemist, Renaissance man, and one of true nice guys of this world, liked to shoot the breeze in this kind of Pacific Rim salon setting. When I was on-island for field visits during the research collaboration with the university, I would join project coordinator Heidel on Honolulu supply runs. As with Colville in 2017, the Hawaii project genetics lived in a DEA-supported locker. In this case, a locked fridge down the hallway from Qing's office.

"3T3-L1," my favorite academic said at one of these teas, setting down his enamel cup and writing the sequence on the dry-erase board beside the break room table piled high with back issues of light reading like the *Journal of Agricultural and Food Chemistry* and *Annual Review of Chemical and Biomolecular Engineering*.[5] "This is a typical lipid cell used in an obesity study. We looked at the impact of a hemp diet on it."

He returned to the table and casually refilled our glasses, as though he weren't about to drop a piece of potentially humanity-altering research he had been quietly conducting in paradise. I mean, if this work comes to fruition, it might alter the very shape of the human form.

"You are skinnier than me," Qing continued, elbowing me impishly. (We're equally thin.) "Not necessarily because your fat cells are fewer. It's because you have less lipid in each of your cells. I have a few more cell blocks in my lipid neighborhood."

You might want to be sitting down for this next part. Qing, stressing that his research is "at a very early stage," said that when hemp is part of a diet, it looks like the 3T3-L1 cells are inhibited. "They appear to stay smaller."

Qing's preliminary research indicates that a hemp diet plays a role in combatting obesity. I realized in a flash that the diet-craze book title is, of course, *The 3T3-L1 Diet: Hemp Keeps You Thin.*

"That research made me interested enough in industrial hemp to want to grow it in the field," Qing went on, breaking out the chocolate wafers that would soon be fighting with my hempseed breakfast over the girth of my lipid cells. "Because the promise is there, I'm willing to spend extra effort to do more studies."

Hemp Keeps You Thin. Ya know, in conjunction with exercise, a comprehensively healthy diet, and genetic good fortune. We now possess, in fat cell blocking, the ultimate diet hook. And the best part is, when the craze hits, it may do some real good, for farmers and customers. I now frequently envision hundreds of millions of people getting visibly thinner over six months or however long by adding a delicious superfood to their waffles.

I see how to get there. Heck, diet books were the rage even before the current obesity and diabetes epidemics, back when folks only *thought* they were overweight, because TV and diet books told them they were. Now they really are, and "food"-related health issues are among the most serious and pervasive problems facing the species.

"3T3-L1," Qing said again as we cleaned up the table several wafers later. Melody and I were picking up some clippers, then heading an hour south to the field while volcanoes rumbled one island over. "Remember that."

"I will," I replied with vim. And I have. Qing explained a game-changing concept magnificently and, equally important for mass media exposure, succinctly. He had me at "fat cells stay smaller."

An Inflamed Species

Even as more research is conducted on the anti-obesity front (obviously no one's really writing a diet book until we see years of double-blind studies, peer-reviewed and all that), we already have the requisite sequel in the *Hemp Keeps You Thin* series lined up. It comes from existing knowledge about another one of hemp's nutritive components: anti-inflammatory properties.

It was hemp food giant Nutiva's 60-year-old founder John Roulac, citing the 2002 paper by David Marcus and the impactful Ernest Small that serves as the first epigraph for this chapter, who first clued me in to gamma-linolenic acid (GLA).

"Many folks don't realize how important GLA is, and how hard it is to obtain from food sources," Roulac told me not long after we met a half decade ago. "It's an omega-6 associated with anti-inflammatory properties."

That is likely because of the eicosanoid metabolites (which are building blocks of fatty acids) in GLA, according to Dr. Dylan MacKay, a nutritional biochemist at the University of Manitoba, although he added that more research is needed before there is consensus on this point. And it is under way, including multiple studies showing anti-inflammatory effects of hemp diets in pigs, mice, and guinea pigs.[6]

The anti-inflammatory discussion is a hot one these days because, well, so many folks find their digestive, immune, ocular, and other regulatory systems in a state of cell inflammation. Diet, air and water quality, and pharmaceutical side effects all probably play a role in this disastrous mishmash.

Let's look at digestion. According to the nutritional philosophy espoused by people like physician Andrew Weil and Stanford microbiology and immunology professor Justin Sonnenburg, what's happening in your gut is very similar to what's happening in soil: Your endogenous chemicals and (ideally beneficial) microflora are speaking to the nutrients and (ideally beneficial) microorganisms in your food. Collectively this internal ecosystem is called your *microbiome*.

To encourage microbial balance in the ol' belly, I eat a lot of fermented foods and home-raised goat yogurt. My anti-inflammatory regimen also includes avoiding unnecessary food additives. I read every label before I crunch something, looking out for sketchy ingredients including corn syrup, natural flavors, or autolyzed yeast.

But what about those who are in a perpetual crisis mode with one of their body's regulatory systems, especially when inflammation might be the culprit (or at least one of the suspects)? GLA's documented properties in this area make it a de rigueur part of a daily human diet. So our sequel best-seller title, hopefully while the original obesity-themed selection is still topping the lists, is *Hemp Keeps You in Balance*.

Hemp Animal Feed

There's one more dietary area that itself might well demand a few hundred million hemp acres. Its manifestation in book form can be called, *Hemp: It's Not Just for Humans Anymore*. I can attest that hemp feed results in maximum performance for my goat herd. But I feel obligated to add that it does *not* mean hemp makes your goats behave. Still, we need this livestock side of the dietary revolution—it is coming, and it will be big. Millions-of-acres big.

Factory farming of 65 billion critters in concentrated animal feeding operations (CAFOs) contributes 18 percent of worldwide greenhouse emissions, including 37 percent of methane, according to the United Nations Food and Agriculture Organization. This is why you roll up your windows when you drive past one of those "ranches."

I was thinking about this after the 2018 Oregon harvest as Edgar, Margaret, and I fed the koi a dash of the hempseed they so love. I filmed the ensuing vegan fish feeding frenzy with a submerged GoPro, and was so impressed that I looked up some stats. Animal feed is a $297 billion industry in the United States, according to the Institute for Feed Education and Research, with 75 percent of that GMO corn- and soy-based (for the moment).

Hemp as livestock feed has a long legacy in the United States. At a book signing following the first talk I gave when *Hemp Bound* came out in 2014 (this was in Denver and televised on C-SPAN), a sturdy octogenarian rancher from Nebraska named Nan approached me and filled me in on a bit of American agricultural history.

"Growing up, I knew it was spring when my daddy had us plant hemp along the irrigation ditch," she said. "The roots helped the ditch walls hold fast whether we faced drought or flood. And I knew it was fall when we set the cattle on that hemp to finish 'em. They loved it."

I can relate—my goats do this sort of joyful sideways hind leg kick when I bust out the hemp-protein meal. We humans should kick, too, as we benefit from healthier, hemp-fed animals: In *Hemp Bound*, we visited with Canadian researchers who had just completed a study showing that hemp-fed hens produced more nutritious eggs than corn-fed hens.

Yet in the United States, hemp as livestock feed is still, astoundingly, fighting for the necessary federal designation known as GRAS (generally recognized as safe).[7] But don't worry: Iowa farmer Ethan Vorhes and the Hemp Feed Coalition team are fighting a disciplined fight on the livestock front. Ethan just wants the competitive advantage he believes hemp will provide his cattle. Today you can feed hemp to your own animals, of course, and to commercial livestock that humans don't eat, like horses and dogs. But not yet to commercial cattle, chickens, and the rest.

When we inevitably win that battle, the numbers we add to the 944,000 jobs that the Institute for Feed Education and Research says the animal feed industry already provides might help raise our percentage of Americans farming close to our target of 30 percent.

One other note on the protein-rich meal that is hempseed oil's by-product: It might provide a competitive edge to nonhuman athletes. It was my colleague Marty Phipps of Old Dominion Hemp in Virginia who convinced me to do the original Samurai protein test that came out at 31 percent, because the Olympic-level dressage horse owners to whom he already provided hemp hurd for bedding were interested in the known nutritional benefits of high-protein feed supplements—from any source.

These are high-performance animals, of the "this filly is worth more than your ranch and mine put together" class. Their owners are looking for the same kind of hundredths-of-a-second performance advantages that human Olympic sprinters seek. So now Marty and I are going to work together to expand this market. Others have started looking livestock-ward as well. Hemp dog treats are already big. Look for hemp-based high-performance livestock supplements to become another deserved craze in the near future.

We need that craze. Both Chad Rosen and Roger Gussiaas tell me that protein meal is the more difficult seed-based product to move in their current inventory. The oil flies off the shelves. My prediction is that protein meal will catch up when hemp livestock feed comes online.

To hold up my end on the human consumption side, I'll make sure *Hemp Keeps You Thin* contains plenty of recipes for hemp protein in bread, cakes, cookies, healthy shakes, and, of course, waffles. All I ask of you is that you grow some of those 234.7 million acres of it every year. Or help ensure that those who do make very fine livings. Which is really the key remaining piece of this puzzle. The hemp diet craze only works if farmers prosper. I mean lucrative, dentist-level livings for millions of farming families.

Farmers in the American heartland and similar breadbaskets around the world know how to cultivate grain. And they do it on a vast scale. It's not uncommon for one farm family to plant five or more sections. Which leads to a legitimate question some macroeconomist might ask about mass hemp acreage: "If we really grow millions of acres of hemp, won't the market crash?" In other words, in our "go big with hempseed" thesis, aren't we violating Wendell Berry's prime directive?

The answer, in my view, is a firm no. That is, if we're smart about developing long-term markets. The surplus value of organic certification plays an important role.

"We offer two and a half times the price for organic on our seed contracts," Roger Gussiaas told me of his Healthy Oilseeds business, where every acre is grown outdoors and the presses never stop.

The 58-year-old Roger, a generous member of Team Hemp who even invites potential competitors to facility tours in the name of expanding the food side of the industry, has a farmer-friendly "act of God" clause with his farmers. This assures farmers some payment even if a crop doesn't pan out. Needless to say, such protections should be a component of any Digital Age Homesteading Act.

If organic hempseed at this scale brings in a conservative $1 per pound wholesale to a farmer who has low overhead on 3,000 acres and who harvests 1,000 pounds per acre, that means a $3 million crop for that farming family. Plus the family's work is done when the seed is delivered to the grain bin. No barn ice-skating, for better or for worse. Although nothing's stopping that family from also marketing an artisan secondary-market product.

It's Raining Organic Hemp Energy Bars

Along those lines, I wanted to check out if value-added hempseed product lines are an option. This investigation gained some urgency when I got an email from a Pennsylvania farmer in 2018 who was having trouble moving her wholesale seed. As markets stand now, if you start working on it with the same vigor we're been discussing on the flower side, you've got a shot. But it's no fait accompli, even according to one of the most successful value-added seed purveyors in the business.

When I first met Evo Hemp cofounder Ari Sherman at the company's Boulder, Colorado, facility in 2016, he and cofounder Jourdan Samel (biz school buddies, both then under 30) were already moving close to a million units a year. These were all-organic, US-sourced hemp energy bars. Varieties at the time included cashew cacao and mango macadamia.

A million hemp bars! That sounded mighty impressive to me, a fellow who had not yet expanded to a thousand units. In fact, you could see the Boulder Whole Foods Market location that Evo Hemp supplied right from the company's front door.

"That's actually the store where I got the original ingredients when we were toying around with our first bars," Ari told me as we toured a warehouse full of ready-to-ship cases of hemp bars. "I bought their bulk hempseed and fruit. Now we sell them the products."

Foolish, naïve me. I thought selling a million products meant you'd cleared a million dollars. I thought these guys had made it. They had been written up in *Forbes*. Then, as we tasted samples of the new mocha chip bars, Ari spilled the beans about how the big-league retail food business works, from the independent entrepreneur point of view. Even at a million units sold, he and Jourdan hadn't been paid yet.

"The way it works is, you get the honor of being given shelf space at a big store," he explained. "But it'd better sell. If it doesn't sell well, you don't necessarily get paid right from the start. The chains set the rules, until you're making them a lot of money."

"Isn't that theft?" I asked.

Ari looked at me for a moment, then tilted his head in a sort of, *I guess I never looked at it that way* kind of way.

Fast-forward three years, when Ari and I spoke again in 2019. I asked first off if he felt Evo Hemp was over the hump, from a bootstrapping entrepreneurial standpoint.

"As most entrepreneurs learn quickly," he said. "There are many different plateaus that an enterprise reaches. I don't think we're ever over every hump. People like us have a hungry nature. But the short answer is that yes, we're over the hump that we discussed in 2016. As things grow, you start to see other hurdles in front of you, the next plateau. Each of which brings its own set of challenges and opportunities."

I asked the obvious: "What's the next plateau?"

"Jourdan and I have had to wear every hat up until now," Ari said. "We're the packing guys, salesmen, we're physically making products, doing product development."

"I've even seen you as chief spokesmodel at trade show booths," I said.

He laughed. "Exactly. The next steps are building out teams. We've hired a graphic designer, for instance, which will free us on the marketing side."

"How many Evo Hemp Bars get eaten every year now?" I asked.

"We'll sell more than five million bars in 2019," Ari said. "Besides Whole Foods, now we're in Costco, Kroger, Barnes and Noble, Fresh Market, TJ Maxx, Marshalls, and Ross."

"As a result, on the manufacturing side," he continued, "our demand has exceeded our supply capabilities here in Boulder. That's a good problem, but we need two extra production facilities for the bars. And now we're moving into other categories, like our topical CBD line."

"With how many farmers does Evo Hemp contract?" I asked.

"Six farmers on fourteen hundred acres paid two dollars per pound for organic in 2018," Ari said.

No point dancing around the bottom line. "Are you and Jourdan getting paid yet?" I asked.

"Yes," Ari said with a chuckle. "Not any crazy executive salaries. But we are able to live a sustainable lifestyle in Boulder. Which is not a super affordable area."

We spent the last bit of our chat commiserating on the "Oh well, we entrepreneurs will sleep when we're dead" theme.

"The work is never-ending, and taxing on Jourdan and me as individuals," Ari said. "We have a sort of positive feedback loop with supportive colleagues and family, which helps."

"I love seeing a hemp entrepreneur's social media post from a forest or a beach when I know he's been working hard for a long time," I said. "I think, 'Good for you. A little vaycay.'"

"I don't know what that is," Ari said.

Yeah, I thought. *And you don't have kids yet.*

CHAPTER TWELVE

Rebirth of the
Biomaterials Era

*For a long period of time there was much speculation
and controversy about where all the so called "missing
matter" in the universe had got to. . . . When eventually
tracked down it turned out to be all the stuff which the
equipment [to study the problem] had been packed in.*

—Douglas Adams, *Mostly Harmless*

When the homeschool lessons are done here on the ranch, we sometimes tune into the terrific Canadian TV show *How It's Made.* In 5-minute chunks, each episode reveals how the sausage is made in the case of everything from marbles to golf clubs to, well, sausages. The ball bearings episode got me thinking about another category of hempseed applications that might necessitate a few million more acres: the industrial. In particular, the components of the lubricants, glues, epoxies, inks, dyes, and paints that so often frighten me when I learn How They're Made.

You see, midway through too many episodes of the show, there's this "nothing to see here" moment where the narrator glosses over a "chemical soak" or "solvent bath" stage that causes me, as a father, to ask, "What chemical? What solvent? My kid might put that spoon / Matchbox car / packing box /mattress / drone controller in his mouth. And don't even get me started on what my goats might do with it." In the ball bearings episode, the line that disturbed me was "An abrasive stone coats the [bearing] with thick oil . . . and it goes into a solvent bath."

Now, I recognize that gadgets require inputs. And I love writing this book on my trusty old laptop. Even Alexander Graham Bell's telephone relied on vibrations in an "acid bath." The key is whether it's citric acid from an organic lime or poisonous hydrochloric acid. I'd like to put forth the proposition that the survival of humanity might depend on feeling good about the explanation for the kind of solvent bath in which our industrial materials spend the night soaking. And that's where hempseed oil and other biomaterials reenter the picture.

Yes, reenter. For the moment biomaterials are still considered "alternative." For most of history, they were routine. Hemp, like a lot of plants, has been employed in industrial capacities for millennia. Before Whole Foods, Home Depot, and that first Pennsylvania oil strike in 1859, that's all there was: what the soil gave, give or take some whale oil. Feeding our industrial pipeline with regenerative biomaterials is really just a return to the way things have always been.

Both hempseed oil and hemp fiber were (for better or for worse) routine components of European age of exploration ship repair.[1] Dutch vessels, upon safely reaching dry dock at Rotterdam, would immediately find their more severe wear-and-tear cracks stuffed with hemp's long-bast fiber. After which, hempseed oil–based sealant (a "chemical bath" about which you could feel good) was used to waterproof—and salt proof—the hull for another arduous journey pillaging Indonesia or Suriname.

In my region, folks find arrowheads, affixed with pine sap and yucca fiber, ready-to-shoot after 700 years. My sketchily glued stereo headphone cushions, by contrast, are warrantied for one year. If *The Graduate* were made today, Mr. McGuire would advise Ben to get into "extended service agreements." Yet I've been so well trained to think that industrial-scale stuff must derive from synthetic sources that any exception still surprises me. Meat tenderizer is made from papaya? Wow. I'd have thought some horrible chemical softener. And it seems downright magical to me that I was able to 3-D-print the plastic goat made from US-grown hemp I mentioned earlier. Not only that: Alongside my colleague Andrew Stoll, I helped make the filament that fed the printer.[2] That goat has become sort of my avatar. I never leave home without it. It's fun to hold tangible evidence of an opportunity to merge humanity's carbon mission with its digital trajectory; to show how a plant becomes an industry.

Hempseed oil and fiber, plant saps, beeswax—they all work so well, every now and then I try to imagine the first fellow to say, "God's world is too messy. Let's make everything in a lab." Turns out "better living through chemistry," a rallying cry of 20th-century industry, wasn't a merit-based philosophy. It was a chemical-industry-based one. In fact, it originated as a DuPont advertising campaign in 1935. And that might've been fine if all the effluent hadn't been dumped into rivers for 100 years.

Patriotic Hemp Paint

In the 1930s, with the petrochemical era barely two decades old, a testy debate raged among fat cats: Do we keep farmers producing our industrial feedstock or do we consciously move to creating it in a test tube? For a century, the farmers and the planet have been losing. The DuPonts and Monsantos of the world have been winning.[3]

But in the same year that "better living through chemistry" entered the national lexicon, not everyone was drinking the synthetics Kool-Aid. So effective was hemp as an industrial component that in August of 1935, the director of research for the Sherwin-Williams Company, C. D. Holley, got fed up with his problems obtaining it. We know this because he exchanged a series of letters on the subject with cannabis prohibition villain Harry Anslinger, director of the new Federal Bureau of Narcotics. In them, he argued that dialing back the reefer madness rhetoric would help his company's bottom line. Holley tried repeatedly, and with great initial restraint, to explain to Anslinger that hempseed oil was needed in the company's paints.[4]

"(Hemp) is an excellent opportunity" to supplement Sherwin-Williams's paint-drying oils, he wrote in one of his letters. At one point Holley pointed out that Anslinger's senseless anti-cannabis policies had forced the importation of 54 percent of the US industrial hempseed oil supply. Hempseed, of course, has no psychoactive properties. He even tried out a farmers-first argument on his bureaucratic nemesis: "It would seem desirable in every way to give the farmers this opportunity [to cultivate hemp for industrial applications, and thus] eliminate the large and increasing import of hemp seed."

That bit of verbal swashbuckling would be funny if it didn't capture the start of cannabis prohibition's acute damage to the economy, especially the rural economy. We've since lost 90 percent of US farmers. Federal cannabis prohibition began two years after the Holley-Anslinger exchange, and continued for 77 years, until the passage of the 2014 farm bill. Since then, hemp has contributed a conservative $3 billion to the economy, based on retail sales reported by Vote Hemp.

Unlike Holley, I didn't need to argue with Anslinger's bureaucratic descendants.[5] For years, hempseed oil has been my go-to mechanical lubricant. I just employed it, for instance, to oil my razor (in preparation for my very occasional facial pruning), and to lubricate my camping stove pump mechanism. My kids and I are, concurrently, running some tests to see how hempseed oil performs as a wood-weathering treatment, versus boiled linseed oil (derived from another biomaterials source, flax). This is for the frame supporting our season-long hemp-field time-lapse video project. If the hempseed oil wins out, we'll use it to seal our under-construction tree house.

My sweetheart, as I read her this section, reminds me that she found our raw, unfiltered hempseed oil a bit gummy as a sewing machine oil. That encourages me to ask one of you ingenious farmer-entrepreneurs to cultivate a few million acres of a variety less *robusto* than the nutritive Samurai she was using: something intentionally clearer, less green, with a more industrial viscosity, for uses ranging from motor oil to rocket fuel. Might partly be a matter of processing levels and filtration.

Compostable Supercapacitors

A word, though, on hempseed oil as a biofuel. Much as I enjoyed driving in a hemp-powered limo with Bill Althouse while researching *Hemp Bound*, our long-term transportation and energy solutions may well come from hemp fiber, rather than hempseed. Fiber launches us into the post-combustion era when solar-charged electric devices, vehicles (including planes), homes, farms, and the industrial backbone are all running bio-based supercapacitors in their next-generation batteries. Arguably the most exciting hemp app of the past 8,000 years emerged from research done by David Mitlin and his team at the University of Alberta, initially in 2014: They found hemp fibers are proving a superior feedstock to synthetic nanomaterials: slightly faster, much cheaper, and unfathomably cleaner.[6]

Here's the very basic story on supercapacitors. Utilizing sheets of one-molecule-thick carbon called *graphene*, these charge and store energy in a different way than the batteries that, if you are reading this prior to, say, 2022, are in your flashlight, golf cart, or solar battery system. And they do it magnitudes faster.[7] The days of waiting hours for a car or solar-powered home to charge are numbered—we're moving to this happening in minutes. The problems are the cost and/or nonregenerative-ness of early sources of graphene.

Stacked sheets of graphene provide the material used for superca-pacitors. What Mitlin and his graduate and postdoc students discovered was this: The surface area of the hemp carbon molecule when stacked at the nano level is superior to any other known graphene source. That's because hemp's molecular arrangement has a unique (yes!) wafflelike formation. Just as important, the cost of the process to get to nano sheets from the source feedstock (hemp fiber) is very low.

"The structure of the precursor turned out to be ideal for the final product," he told me when we spoke in 2015. "The cost came down to a buck a kilo to make [from $1,000 for then-existing graphene options]. It is comparable with commercial graphite sources for battery applications. It's the same capacitor in a lithium-ion battery, but with a better spin on it. In the end, it helps make a better battery."

How on earth did Mitlin even think to examine hemp for these amazing properties? Well, you have to love it when a study is co-funded by the Canadian Institute for Nanotechnology and the Alberta Live-stock and Meat agency.

"Hemp was very available in Alberta," Mitlin said. "[Provincial funders] were looking for fiber applications."

Not that he didn't try other biomaterials in his search for superior performance. Evoking Edison, he told me with a smirk, "We tried egg-shells, even banana peels." Talk about plan B.

It's hard to imagine a more vitally disruptive technology. Conventional batteries are an environmental disaster. And not just the hundreds of millions of Energizers that go into the world's landfills. I can confess this battery quagmire applies even on my green-sounding solar-paneled ranch. The dozen lead-acid golf cart batteries that store the sun's energy for my laptop, for instance, are an embarrassment (lead mining can be terribly polluting), and the rare earth minerals in our current generation of digital device batteries require both an unacceptable environmental and human rights cost.[8]

One could argue that hemp's ideal nanostructure is just another example of humans and the cannabis plant co-evolving. But in this case, I don't know how the plant knew in advance to evolve into that perfect folded pattern at the molecular level just in case one day its friends the naked apes developed supercapacitors. So I'm going with, "Hallelujah!"

If your family, like mine, wants to at once raise goats and have Netflix and space travel, we have to as a species move beyond combustion and into a solar-based society and economy. That means something clean and regenerative has to hold the charge that the sun is giving us for free. Graphene has to become biomaterials-based. Synthetic, mined, or cost-prohibitive graphene sources are off the table.

Of course, solar-charged compostable batteries might mean I'll have to start buying farming equipment. Remember the Hana Ranch, leading the way on electric farming? I am so ready to tell Aaron I have abandoned my oxen harvest plans in favor of hemp power.

Perhaps most exciting of all for independent hemp farmer-entrepreneurs who might be interested in providing feedstock for this kind of technology was Mitlin's answer to my last question that September day in 2015. We were at a convention in Lexington, Kentucky, and I was fresh from a nearby harvest (on a field at the Wendell Berry–affiliated St. Catharine College). I showed him the raw hempen bracelet I'm always wearing for the several months it lasts following such harvests. It had chunks of both bast and hurd on it. I said, "Would this work for a supercapacitor application? Does it have to be prima donna textile-grade fiber?"

The answer was a definite no. "The stuff we took was from a pile the size of a house in a storage bay," Mitlin said. "We put it through washing steps before we worked with it, but it wasn't laboratory-grade material—someone took a shovel and put it in a bag."

That's good news for a coalition of hemp farmers just launching a cooperative. And it was to supercapacitor research that Wild Bill Billings donated his first hemp-fiber harvest in 2014.

Now, the gap between a great idea and market acceptance can be years. During a transition period from petroleum, sure, liquid hemp power at the pump might be a piece of the puzzle that keeps our millions of acres planted and our legacy fleet of diesel vehicles on the road. And the seed oil will, let us hope, always be lubricating our tools, gears, and axles as Charlie did with our seed drill. But real transportation victory comes when all our cars, trucks, and planes are powered by rechargeable and recyclable hemp-based batteries. You'll know this has happened because, while the speed limit will have gone up, life will have gotten much quieter.

Just as hempseed oil is now Charlie's and my favorite ranch lubricant, so I'd love to see all our industrial solvents, glues, and binders become plant-based, renewable, and safe once again. I mean, is anyone loyal to the terrifying epoxy polymers currently fusing the parts of their washing machine?[9] How much would you love regenerative duct tape? If you're thinking of starting a company to accomplish some of these goals, you might consider naming it after C. D. Holley.

For the regenerative entrepreneur, the health of customers is a key component of the brand. That's one reason I'm paying through the nose for the new run of compostable Hemp in Hemp labels, just to walk the walk. Their nontoxic adhesive, in particular, feels like a step in the right direction. Think of the sticker glue on nearly every apple you eat. How's It Made? It's time to ask.

The biomaterials train has left the station. If you're a customer, it's up to you to hop on board and make the resulting markets stick. Then the carbon keeps getting sequestered. Plugging those How It's Made gaps in our wider industrial processes is a vital and long-term project. Agriculture can show the way, but it's about the whole society getting on board, in all sectors, on all cylinders. Every industry, from energy to transportation to manufacturing, must transform to a regenerative mode. A factory, glue product, or road might look nearly the same as it always has, but there will be two differences: The bio-based one will perform better, and will now play a role in saving humanity rather than destroying it.

Plus we'll get better industrial advertising slogans too. Rather than "better living through chemistry," our style of tagline will be superimposed over images of reusable hemp-built rocket boosters sticking a landing, happy families loading up their electric car with a picnic basket, and an auto mechanic lubricating an axle with a tube of clean, safe plant-based oil. One slogan could read, simply: "Hemp: Superfood for Our Stuff."

Longtime food industry professionals tell me they believe farmers will be receptive to human, livestock, and industrial seed-oil and fiber applications. Steve McGarrah, founder and CEO of the Kansas City–based High Plains Nutrition animal feed business, describes the heartland farming situation starkly. "We've got upwards of a million starving Midwest farmers losing hundreds of dollars per acre on soy," he said. "I estimate twelve million acres is needed just for animal feed. It's such a fine match of need and timing that I've considered paying to provide farmers with their initial seed."

He's not waiting for the Digital Age Homesteading Act. The plant, in alliance with farmers who grow it and the engineers who figure out how to make it into useful stuff, is ready.

No More Plastic Crap

Supercapacitor scrap piles aside, the fiber side of hemp is hard. It is physically hard, which is good—it's durable, sturdy, pound-for-pound rivaling spiderwebs for the title of world's strongest material. But it's hard for independent farmers to market, because it requires immense volume to feed production-scale processing. And yet we need it. If seed oil can help with our resins and epoxies, it's the fiber that will compose the actual stuff in our regenerative future.

I had no choice but to work on this personally. About the time we had our second (human) kid, my sweetheart officially instituted the No More Plastic Crap rule that still governs our lives here on the Funky

Butte Ranch. We were sick of the Great Pacific garbage patch, then the size of Texas, now the size of Russia. Plus, our 1,200-square-foot adobe house has only so many molecules of airspace in it, for any materials, regenerative or otherwise. When I don't wear my slippers for a simple trip from office to bathroom, I severely injure a toe.

The final straw, though, I'm pretty sure, was the mountain of off-gassing, offshore-produced, already-breaking gear we compiled from well-meaning kin when the first child arrived. Plastic swings, leaky sippy cups, pesticide-soaked onesies made by other kids in Indonesia. At least the diapers were hemp, for both kiddos.

Kicking petrochemical plastic is, of course, a sound policy, and I say that only partly for reasons that anyone with a strong, treasured life partner will understand. The least I can do is follow the lead of the woman who makes my hemp underwear. She makes me feel like Pa Ingalls without the cultural insensitivity and the tobacco addiction. But also she's right: Decaying plastic particulates really are bad for you—did you know if you use petro-plastic bottles, estrogen-like chemicals can leach in your drinking water? Well, only in 70 percent of samples in one study.[10]

Even the everyday devices we press against our bodies contain unnecessary dangers. In 2018, Samsung's president Kim Ki-nam apologized (after 10 years of stonewalling) to the dozens of employees sickened in a company cell phone factory from the routine toxins that we, in turn, carry with us everywhere. "Our tools are killing us," the soil expert Ray Archuleta says.

It's not that inconvenient to give up petro-plastics and the chemicals that go into making them. Really the only substantive lifestyle change I notice since the "no new plastic" rule went into effect is in my family's "to bring" lists for our very occasional town trips: hemp grocery bags instead of "plastic or paper" in the store, for instance, and wax paper to wrap bakery bread.

At home, the only adjustment I have to make is accepting the dangerous potential in all our repurposed glass bottles and containers. Half awake at breakfast time, if I grab a jug with white powder labeled COCONUT FLOUR? Might be that, might be salt, might be diatoms. The other day I heard myself asking my oldest son, "Would you please bring me my water glass? It's the spicy pickle jar."

One has to repurpose, though. Our devices, during this brief, ridiculous planned obsolescence era, are a key problem. A 2016 United Nations study reported that 44.7 million metric tons of electronic gear is dumped planetwide each year. Yay, people. We Americans, like folks in most of the world, just throw out 80 percent of our old phones and other digital widgets. Japan's a bit better, with store-based recycling programs that get as much as 50 percent of old gizmos.[11] We have to figure out a way for our stuff to be compostable, or at least genuinely recyclable. End of life for a product is as important a stage as the day we buy it. Hemp fiber, wellspring for the coming worldwide return to biomaterials feedstock for our entire industrial pipeline, is on the case.

A decade ago I wrote a book, *Farewell, My Subaru*, about greatly reducing my fossil fuel use while keeping digital-age comforts. I bolted some solar panels to the roof, started eating more ranch-raised food, and began driving on vegetable oil, those sorts of things. Even though I calculated that I reduced my fossil fuel consumption close to 80 percent as a result, I still consume more fossil-based energy than the average Filipino village. And even with the dramatic reduction in new plastic purchases my sweetheart has instituted, what seems to stay in our life no matter what we do is the dang wrapping on all our widgets and even our food.

Actually the problem is the wrapping for the wrapping: Even when a package comes in compostable cardboard, it so often is shrink-wrapped in hellish clear plastic. Gobs of it accumulate in our home, edging the dogs out of their beds. This has become a top fiber app mission for me: I would so love hemp to shake up the packaging industry.

A solution was handed to me in the course of reporting on Margaret Flewellen's Oregon products, specifically her Zenith CBD hemparettes. Until you get to each carton's "overwrap" (the industry term for the clear stuff that drives me mad), Margaret's completely vertically integrated product is the poster child for the regenerative packaging effort. First off, a Zenith, with hemp grown by Edgar, is wrapped in organic hemp rolling paper with hemp filters. A dozen are packaged in a box made entirely from hemp by a Eugene, Oregon, company called Hemp Press. So she's supporting on-brand, regional ancillary companies, which is easy when their performance makes the grade.

"It's more durable than the tree-paper packaging we used at first," Margaret told me when I examined her packaging process. "At scale it still costs more than tree paper—fifty-four cents per box—but it's getting increasingly cost competitive."

Now on to the overwrap for the boxes. Margaret's overwrap machine, a Xopax brand model, is about the size of a printer. She sets a box of Zeniths on it, and it seals it in what I've always called "cellophane," in the name of what our era considers safe delivery.[12] For now, this is usually made from petro-plastic. When I told her that overwrap fills my house and frays my nerves with each crinkly step I take, Margaret nodded empathetically. "The plant-based wrapping is coming," she said. "It's the last piece of the puzzle. You should ask Randy about it."

"Randy?" I asked. "Your uncle?" I loved Randy, I farmed with him often, but he didn't strike me as someone who read biomaterials journals. I had recently watched him, armed with a shotgun and wearing a muscle shirt, chase off two squirrels who were chewing a drip line.

"Ask him," Margaret said, wrapping another box of Zeniths.

On planting day in Oregon, 2018, I did. Turns out Randy Flewellen, bearded, 59, and generally camo-clad, is a genuine thinker. When I started griping about the Overwrap Gap during a seed drill FLOATER delay, he dropped this bombshell.

"Oh, they can make overwrap from biomaterials," Randy said as I handed him the wrong size socket wrench for the jammed seed chute (so close, just not metric). "Already do. Company called AET. Applied Extrusion Technologies."

Needless to say, I could hardly believe my ears. How could he know that?

"How do you know that?" I asked. Doug Adams reminds us never to believe anything we hear at cocktail parties. But I would argue a cocktail party has nothing on a farming delay as a fertile BSing ground. It might have to do with all the actual bull dung around a field.

Randy said, "I was night manager for a printing company called Deluxe Packaging that used their wrapping films." He chucked the wrench in the growing discard pile, bleeding some more on his shirt from the effort. "Some of them were plant based. I worked that shift all night for eight years. You could get bio-based films in all gauges, all sizes and styles, including transparent. The plant stuff was the most durable material I worked with. Man, during heavy production times it could be eighty-, ninety-hour weeks. I wrapped packages till cellophane was in my dreams."

Now it's in mine.

"I think they might've got bought out!" Randy shouted as Edgar fired up the tractor again. "I think the company's called Taghleef these days, out of Canada, or it might be Dubai by now."

It was. So I called 'em. Danged if it isn't true. The product manager for the company's plant-based, flexible film product line, called Nativia, is Emanuela Bardi.[13] Based in the company's Italian offices, she told me via email that the product line's range of flexible films includes a transparent offering, with thicknesses ranging from 17 to 50 microns.

"Nativia is still a niche market compared to the core business of Taghleef at the global level," she wrote. "But sales are growing by double digits each year . . . and there is a growing consumer awareness about more sustainable packaging."

Now, the Nativia line is just one division in one packaging company. And the actual plant feedstock company for Taghleef is owned by ag giant Cargill (as is the corn binder in my hemp plastic goat, until some of you engineering grad students get working on hemp-based binders). So it ain't organic. But it also ain't petrochemicals. Which means it's a start.

So now I'm leaning on Margaret to see if she can't wrap her Zeniths in this stuff, or something like it. Once she does that and her Natural Good Medicines enterprise grows to a massive scale, I'll get on her to make her delivery vehicles electric. This is what I call connecting the dots in our enterprises. And so humanity merges onto the comeback trail.

CHAPTER THIRTEEN

Farmer Fiber Collaboration

Long-lived (ant) colonies in the desert regulate their behavior not to maximize or optimize food intake, but instead to keep going without wasting resources . . . (this) allows the colony to deal with high operating costs. . . . The ants have evolved ways of working together that we haven't yet dreamed of.

—DEBORAH M. GORDON[1]

Packaging materials are an impactful starting point for regenerative enterprises, especially from the entrepreneur's standpoint. As a production-level buyer, you can, as Margaret does, make good decisions in your product line's containers. But how do 20-acre farm-based enterprises make millions of tons of hemp fiber? The answer is, collectively. Our rugged individualists must collaborate.

Easier said than done. An agricultural truism holds that getting farmers to cooperate is like herding cats. It's actually worse than that. Cat herders, commiserating over a cannabinoid beverage at the end of a rough day, describe their shift as "like herding farmers." Rather than attempting to link up when it comes to their proprietary apps on the flower or seed side, it is much more likely to be in a region's farmers' best interest to join together on the fiber side. Precisely because the cats don't need to be herded. For the most part, they can stay on their own turf.

Your hemp fiber is like a mountain: It's there, no matter what your primary hemp application is. And it ain't worth much until you make something from it. One of my consulting clients was offered $250 per ton for what they call *mixed fiber* bales—straight out of the field. That felt low to me. A Louisville, Kentucky–based fiber-processing company, Sunstrand LLC, offered contract farmers $600 per acre for harvested fiber in 2018.[2] Although in and of itself, raw fiber doesn't provide a bonanza, even $400 an acre net isn't bad if you're also deriving a living from the seed or flower sides of the crop (or both).

Better still, when farmers follow Dolly and Wendell's directives, their enterprises' fiber value increases immediately once they make their initially low-value fiber into any finished product. If bagging clean hurd is all you do, as Marty at Old Dominion demonstrates, you quadruple your fiber's value.

If all you have to do is open your gates for your regional fiber co-op's dump truck once per season, and the next you hear about it is when you receive a check for your share of the retail value of a final horse bedding product, fiber might prove a real benefit to independent farmer-entrepreneurs. But even though fiber might be the component of hemp's (and other plants') architecture that will play the largest role in extending human tenure here on terra firma, successful development of biomaterials as a mainstream industrial feedstock is far from guaranteed.

In any promising industry, it's the people already trying to survive in it who will deliver the cheerleader a reality check. There are three common themes in all my interviews with bio-based material engineers: Fiber processing requires pretty large entry costs, massive amounts of material, and, depending on the application, a fair amount of expertise.

For instance, when you ask plant-based materials engineer Patrick Flaherty, founder of Kentucky's PF Design Lab, what an independent farming cooperative might want to know about industrial fiber like plastics or boxes, he pulls out three different composite strips (called *coupons*), flicks them like diving boards, and replies, "Are you interested in strength or stiffness? What's the aspect ratio in your product? What vibration damping properties are you looking for to dissipate energy?"[3]

There's a science to fiber, in other words. *Aspect ratio*, for example, is the length of the fiber pieces within your plastic composite or cardboard divided by its diameter. It will be different in, say, a skateboard versus a compostable shipping box. Fiber pros like Shane Ball, who for two years was farmer relations manager for Sunstrand, said, "Basically we could refine fiber into whatever the app is—we could process to the micron." These guys have engineers and chemists on staff.

For independents, there is a low-hanging fruit on the fiber side, and it's that hurd: Our inner core of the hemp stalk once it is separated from the long strands of outer bast. Hurd is relatively easy to produce, once you have the equipment and sufficient supply. Separate the hurd from the bast, clean and chop it to the right dimensions, and bag it.

Hurd markets are both existing and growing. The most rapidly developing ones are hemp building feedstock and animal bedding. As we discussed, Marty Phipps still had to import hurd to meet demand for his high-end horse bedding as of 2018. Hurd was the fastest-growing segment of the Canadian market in 2016. Which isn't really saying much, since seed-based apps still dominate in Canada and hemp flower wasn't even legal back then. The point is that relatively easy hurd apps like hemp-based building are coming into their own.

But hurd is still a crop. Farming practices affect fiber properties, which then contribute to the properties of the market-ready fiber. Still, the technical challenges for a simple hurd operation aren't a deal breaker for a coalition of independent hemp farmers.

"Hurd is hurd, pretty much," Flaherty said. "But when it comes to [bast] fiber, input quality is critical. And time. It can take three seasons before you dial in your fiber yield and quality protocol. And before you fully learn your decortication equipment."

So our putative fiber cooperative might be wise to invest in a facility and equipment that can handle all sides of the stalk, but start with hurd. Become pros at that part of the plant. At the same time, this shrewd co-op has members working to build mutually beneficial relationships on the retail side (obvious choices are hardware, livestock, and home-supply outlets) while developing a reputation for providing something that is new, interesting, regional, and of good quality. "Nebraska Hemp Fiber Co-op," for instance, is a terrific building material brand.

If the co-op's members also find they can market the lower-quality bast for the local university's battery research, a paper-pulping endeavor, or as spill-cleanup material, that's just more revenue for the co-op. Sunstrand even has customers who use the residual hemp-fiber dust that the company's filters catch for compost.

PureHemp Technology, a Fort Lupton, Colorado, fiber-processing company, works on multiple streams for its fiber output. Its factory floor reminds one of the golden age of analog industry—loud machinery turning biomass into everything from paper towels to lignin-derived sugars.

The next obvious question is the cost of a professional-grade processing facility that can handle both bast and hurd. In the fiber world, said fiber entrepreneur Ryan Doherty, president of Virginia-based Hemp Ventures, this includes decortication equipment for separating hurd from bast, but also a range of other processing machinery should the facility aim to be "turnkey," meaning it's able to process both market-ready hurd and bast products.

When I surveyed materials engineers in the hemp sphere, the price of a high-end, medium-scale processing facility, made by established European agricultural processing companies Van Dommele or La Roche, ranged from $8 to $10 million. On the opposite end of the spectrum, a quick-and-dirty way to get at the hurd only if you don't care too much about uniformity on the hurd side or the bast for anything but pulp or batteries, is a hammer mill. Professional models start in the $10,000 range, and really hard-core bad boys that can handle hemp's strength run about $40,000, not counting facilities and other costs such as baling equipment, storage, and bagging machinery.

But for a moderately priced professional facility that can process a farming community's fiber production pretty much 24-7, Flaherty reps a more modular decortication system from a British company called Tatham. It starts at about $1.5 million, but for true industrial-scale processing of both bast and hurd, he said, it can run several million more, when you figure in equipment, facilities, and consulting expertise.

This facility, which can process up to two tons of fiber per hour, would be in reach of an independent farmer cooperative, especially if a member knows how to write grant applications to cover initial fixed costs. But even then, Flaherty told me, there is a knowledge base required in the field, especially right after harvest (when you're already rushing your seed to be cleaned and dried).

"On the farm, you'd better understand retting," he said.

Retting, as was discussed a bit in *Hemp Bound*, is the ancient art of nurturing a field-side fungal process in the weeks following the fiber harvest. Retting seasons the fiber and loosens the lignin glue that fuses the bast and hurd components of the hemp stalk. That makes decortication much easier. Retting is not super complicated (you have to turn the fiber windrows at appropriate intervals) or very different from the mode you see in medieval French sketches of hemp harvesting. But you have to get it right.

"When it comes to fiber processing," Flaherty said. "Junk in is junk out."

That means knowing when the fiber is just the right golden gray color, but not too yellow-gold. Not too wet, not too dry. And ready for the decorticator.

"You learn to feel when the fiber is ready to leave the field," Flaherty said.

"You can pinch it and feel the fiber and the inner core when it's ready," Ball told me. "We want the moisture at fourteen percent or less before baling. I tell ya, I learn something about fiber every time I head into the field." You and me both, brother.

On the cultivation side, if you're growing for fiber, plant tight—20 or 30 pounds of seed per acre. This is for strong mental health upon harvest and retting: Thinner stalks are easier to harvest. If you want to give a fiber guy like Shane Ball heart trouble, show him a picture of the 15-foot monsters the University of Hawaii project grew in 2018. Beautiful plants, and I loved taking pictures looking like a hobbit next to them. But difficult to bring down with anything short of artillery. Shane, 48, took a physical step back as if punched in the stomach when I proudly shoved my phone in front of his face and said, "Look! A project I'm working on is growing a fiber cultivar!"

Consistency is a factor as well. Let's say your regional farmer co-op plans on investing in a hammer mill, and your product is hempcrete building feedstock that you'll supply to your region's hardware stores. Some hempcrete builders want hurd that sports consistent dimensions and is up to industry standards for cleanliness.

Colorado-based hemp builder John Patterson of Tiny Hemp Houses says that though each structure's needs are distinct, he likes to work with hurd "not exceeding one inch in length (for structures, smaller for plaster)."

The diameter is also important, he told me. "It should be 'split' like firewood down to one-eighth inch or smaller." Not nuclear physics, but there's more to it than simply baling your fiber at harvest, hauling it to your facility, and shredding it in your hammer mill. Some builders tell me that they actually want some bast fiber left in with the hurd for strengthening the building feedstock. Others tell me they like to use hurd churned up to very fine dimensions—almost a powder. So even with simple hurd apps, your processing enterprise is going to need to gain professional expertise quickly.

Which is to say, I see the cautionary point made by the pros in the early fiber world who understand about industrial specs and commodities markets. It can seem a little naïve to the guys with materials engineering degrees to envision a bunch of New Mexico or Minnesota farmers in overalls competing in the big-time fiber market. I suggest we independent entrepreneurs take that as a challenge.

And now for the million-acre question: Let's say you get a bunch of your region's farmers together, form a fiber cooperative, and wrangle the initial costs for your facility. Just how many farmed hemp acres does it take to feed a fiber facility? Depends on its capacity. Flaherty and I did some napkin math together in his booth at a Texas conference in 2019. We were surrounded by hemp surfboards, hemp plate-ware, and some spherical, thermoplastic compounds he's designing for the DIY community. He even had a line of hemp composite guitar amplifier dials. Icing on the cake? Our "napkin" math was done on the back of my hemp business card, made by Colorado's Tree Free Hemp company.

Here's what we came up with. Let's say the Tatham decorticator facility can process a ton of mixed hemp fiber an hour (this is conservative, Flaherty said), or 24 tons per day, or 6,264 tons per 261-day working year. If a hemp field produces two tons of dry mixed fiber per acre (also conservative), that means one facility would require 3,132 acres to run 24-7 during every workday.

That's not a small number. But it's doable. Montana farmers planted 22,000 acres in 2018, according to Vote Hemp. See how we're inching toward those 234.7 million target acres?

One economic reality we have to our advantage is that fiber has a limited range from farm to facility. Conventional wisdom (meaning pretty much everyone you talk to, the world over, who processes fiber) says that the sheer amount of biomass involved in a commercial-level operation generally necessitates cultivation within 50 to 200 miles of a facility. I think that might change when demand for value-added bio-materials products matures, and when a cooperative is able to handle its own harvest deliveries. But for now it means that if you and your hemp colleagues launch a fiber enterprise in your region, you will probably be the first ones.

CHAPTER FOURTEEN

Cleaning Up with Plants

If this [material that never wears out] went on the market, it'd destroy the whole economy.

—MAJOR NELSON to Jeannie, *I Dream of Jeannie*

O h, how easy our seminomadic ancestors had it with hemp. No composite strength tests or micron measurements. Just food (seed), shelter and clothing (fiber), and flower for get-togethers. Ya know, tri-cropping. But if you can get a professional-scale fiber project off the ground, bless you. For what? For the carbon sequestration you're facilitating given the vast acreage necessary for a viable operation. You might even call your hurd product "Carbon Sink Hemp."

In addition to livestock bedding and construction, a hemp fiber sector that's about to explode is phytoremediation (soil building). Since both construction and phytoremediation are concerned with cleaning up past messes (cement plants alone contribute 5 percent of worldwide carbon dioxide emissions, and at least one-third of the earth's soil needs some form of healing), here are a few more things to know about both.[1]

Carbon-Negative Living

Though we've been calling it *hempcrete*, a term conjuring images of heavy blocks, hemp building is already proving to be extremely varied: Hurd is being used for structural building, for insulation, paneling, and flooring.[2] I've even seen engineers at Canada's Composites Innovation Centre make a functional sound barrier wall out of it.

When we speak of hempcrete, it's shorthand for hemp hurd mixed with lime (or other binder) and water.[3] But we might see other modes of utilizing hemp for building, ranging from hemp-based spray insulation to digitally printed structures. In these cases, a key question to address is, "What are the binders and sealants?" The goal, remember, is not to be scared of the chemical-bath step in any industrial process.

The early modern hempcrete structures I've explored have all been highly functional—they can be virtually fireproof (blowtorching hemp-crete videos are all over the net), and great for folks with chemical allergies. In Larimer County, Colorado, Melissa and Josh Rabe built (I mean personally built) a 1,500-square-foot hempcrete cabin at 6,200-foot elevation that I entered in a freezing rain in 2015. Inside, with no fire, it was toasty.

Part of that has to do with their attention to detail: There were no doorframe gaps, for instance. A lot of the R-value in any structure can depend on the quality of the build. But when done right, hemp building not only breathes, it regulates heat very well besides. The Rabes even stuccoed the cabin's interior with hemp-cased plaster.

Natural building with hemp is also much less energy intensive during the actual build than concrete, drywall, and stick-frame building. Hemp-crete doesn't require the intense mixing heat that concrete does, for instance. And if the hemp feedstock is grown regionally, transportation energy use is much lower than drywall and lumber shipped across an ocean.

Hemp-built structures even continue working on climate change after you move in. "What is happening as the lime and hemp hurd fuse is they are capturing carbon," said Patterson of Tiny Hemp Houses. If anything, Patterson explained, your walls are hardening as the hurd, in conjunction with its hydrolyzed lime, sequesters carbon.

The hempcrete industry was embryonic when I wrote *Hemp Bound* in 2013—there were perhaps a dozen hemp homes in North America. Today there is probably a hemp building specialist in your state. Indeed, last month I got cleaned out in a poker game by two-time NBA Hall of Famer Don Nelson at a gaming table set on a hemp floor in his hemp-crete house in Hawaii.

It's lovely that hemp hurd is in demand for building—that justifies investing in these expensive processing facilities—but it's even better that it works. I knew hemp building was for real the moment my sons and I made a hempcrete ball in about 10 minutes. We used hemp we grew in Vermont, and lime sourced regionally. For three years that ball has traveled with me, in no special packaging, to nine states, three countries, and four tropical islands. Now, anytime I see someone building a green business storefront with pressboard, I think, *What other species constructs its shelter out of things poisonous to it?*

Despite its growth curve, hemp building is still in its early developmental stages. There is no one type of hemp structure—they can be as varied as the owner's needs. Load-bearing structures will comprise different hurd-lime recipes than hemp insulation.[4] This is why Sunstrand talks about offering fiber options to the micron. And here in the US Southwest, where I live, the adobe-like mixture will be different from the Manitoba virtual igloos that I profiled in *Hemp Bound*.

Set (or hardening) times for each layer of a structure will vary by climate as well, according to Patterson. It even matters what the weather is like on each morning of a build. Which kind of gets at the root of the issue with regenerative building: It's not plug and play. And its implementation involves a lot of building code issues, which vary per region. We might not be able to (nor do we necessarily want to) map the Home Depot / subdivision mentality onto the biomaterials construction era.

Which makes it ideal for artisan builders such as the folks at Colorado's Tiny Hemp Houses and Left Hand Hemp, Mexico's Terrenos Tulum, Hawaii's George Rixey, Idaho's Hempitecture, Ukraine's Hempire, and the pioneers Steve Allin and Hemp Technologies. It also means a hurd provider is going to have to listen to customer needs.

It's not that hempcrete building, at its core, is that difficult. In fact, my job at the builds I've visited has been simple: just tamping the mixture into the hollow wood frames that encircle the rising structure. In the ideal situation, when you finish a two-foot-high layer of the structure, it has set sufficiently to allow you to move the molds on top of that level and begin tamping the next one.

The best and the most challenging realities embedded in hempcrete reside in its versatility. Each build will demand distinct ratios of the three basic ingredients. That soundproof wall I tested in Canada worked because the engineers jacked up the hurd ratio, leaving fewer of the air pockets that a builder might desire for insulating qualities. As a result, you could barely hear a roaring air compressor on the far side of that wall. Useful for folks who live near highways.

When it comes to those folks exploring the idea of 3-D-printed homes, or spraying building fiber out of hoses using the liquefied cellulosic components of the plant, these have promise, in the right situations, such as in or near urban centers. In the more remote, rural spots on the planet, though, a common problem with high-tech solutions to age-old problems is that when the tech side of the operation needs repair, the installers are usually long gone. That's one reason I like solutions that can be done by hand or at least by accessible mechanical means. Even if it requires more brain power and muscle power on-site.

I do think 3-D printing of industrial components from hemp and other biomaterials will play a role in humanity's near-term future, in applications such as body-friendly medical devices, regenerative roads, water filters, and compostable serving ware. Pulping biofiber for consumer items is not immensely difficult or expensive. Imagine every "disposable" cup and plastic fork suddenly winding up in garden soil instead of landfills. You could print your personally engraved coffee cup at the café, then get a dollar off every time you reuse it.

Hawaii, because of its remoteness, is the poster child for industrial independence: Everything not grown locally has to come from far away. The Aloha State will benefit when a \$12 car part can be grown and printed instantly instead of hauled via diesel tanker halfway across the world. How fun it will be to reply, when asked at a cocktail party what you do for a living, "My co-op's crushing it in regenerative toilet paper, truck bumpers, and lunch trays." I'll start gaining real confidence in the wider economic momentum shift to farm-grown industrial feedstock when both the rockets that SpaceX sends forth and the power cells that provide their energy are made from biomaterials.

It's important to remember that hemp fiber can be a major contributor to the biomaterials feedstock effort, but it's far from the only useful plant. This is especially the case when it comes to "waste" cellulosic material (fiber) for use in pulping applications like lunch trays, folding chairs, and serving ware, and in biomass energy. In *Hemp Bound*, we visited the Feldheim region of Germany, where municipal trucks pick up waste biomass from all the region's farms to use in a gasification energy facility owned by the local agriculture cooperative. The region ended unemployment and gained energy independence by making use of what had once been considered trash.

I learned a lot about the broad array of helpful plants, algae, and fungi standing by for us out there in the soil, ready to join our biomaterials arsenal, when a European fiber salesman opened a briefcase in front of me at a Canadian convention. He looked like nothing so much as a hot watch seller in an alley.

The contents of the briefcase blew me away. He showed me about 40 different fiber swaths, including flax, kenaf, kapok, sisal, ramie, and vetiver, each with its own set of pros and cons: tensile strength, ease of cultivation, rate of degradation, fiber density. And all of which he could get for me in industrial quantities at low, low prices if I acted fast.

Which is good to know, because the real-world industrial marketplace has an amazing diversity of needs. If we think of just, say, shipment packaging, there is a wide range of strength, thickness, and stiffness that folks seek. Plus, we're surely on the cusp of some new plant-based breakthrough. Next year's killer app might prove to be an obvious-in-hindsight distillation of fiber lignin that provides the final demise of the cancer cell. Or maybe ground-seed-hull paste will reverse bee colony-collapse disorder. Maybe hurd, when mixed with tungsten and quinoa husk, will allow lightspeed travel or at least teleportation. After hemp's ideal molecular-supercapacitor evolution, nothing going on inside a plant's brain will surprise me.

Healing Soil

The other important hemp-fiber market that could very well prove to be a humanity saver involves doing nothing with the harvest: just planting and letting the hemp do its thing. And that market is phytoremediation. Ninety-three percent of US corn and soy acreage is planted with GMO agriculture (which only started in 1996, by the way), and 96 percent of

India's cotton acreage is GMO. This means that at least that percentage has an herbicide or pesticide such as glyphosate, 2,4-D, or atrazine to be cleaned up. Environmental remediation, in other words, is a major coming agricultural application in itself. In fact, it's an industry projected to be worth $122 billion annually by 2022.[5]

I've gotten calls from folks at two different entities who work in environmental consulting, wanting to see where hemp might fit into their existing work cleaning up Superfund sites.[6] A New Mexico State University researcher is using the Samurai cultivar in a mining-soil cleanup study as we speak. And hemp does have some existing research behind it as a soil cleaner. This is where Professor Qing Li comes in again.

In another preliminary study, Qing and his team found that 80 percent of atrazine in soil was removed 50 days after planting hemp. Further along is his team's published study showing that hemp reduced the levels of chrysene, a polycyclic aromatic hydrocarbon (PAH) and suspected carcinogen found in coal tar and tobacco smoke, in soil after 28 days. PAHs, Qing's paper asserts, "are often the most hazardous components of oil spills."[7]

Folks who love hemp sometimes like to think that there is some magical component embedded in hemp's biochemistry that makes it such a superlative phytoremediator. Qing is careful not to overstate the case.

"I think that people will not argue that hemp is a phyto plant," he told me in a 2019 interview. "Why I can say that is that many plants can do the job: Chemicals will degrade, because plants will host microbes, and microbes will do their job. Is hemp better than others? Well, that is what we are interested in studying. The root is massive, which may prove beneficial."

Don't I know it. As a fellow who hand-harvested two crops in 2018, I'm well acquainted with how impressively strong those roots can be. But I'm so thankful for them—they are the architects of the underground ecosystem that we're endeavoring to build: The fungi, bacteria, and the rest of our invisible friends appreciate the room to grow in their soil-based condo complex. It's good to have friends in low places.

And please remember, if you aim to grow hemp and have any question about the history of your soil, do a baseline pesticide, herbicide, and heavy metals test before you plant. You might have to find more than one testing lab—one for nutrients, one for heavy metals, and so on. Equally important, please remember to only plant any crop intended for food if your soil is clean in these areas, and is at least three years from most recent application of an herbicide such as glyphosate or atrazine.

If not, plant hemp anyway. Just don't eat it, or let your livestock eat it, or sell it to others. You can make this soil cleaning your business: US farmers know that much of their farmland is sick—they have to dump more than twice as much nitrogen on each year as they did in 1964.[8] Europe has 3 million contaminated sites, the United States has 1,300 Superfund sites, and China estimates that 20 percent of its soil is polluted.[9]

Wanting to be clear about the importance of remediating-before-selling, I hesitate to mention this somewhat jaw-dropping postscript to Qing's preliminary inquiry into hemp's phytoremediative qualities. But he did tell me that it appears that hemp might in certain cases render some petroleum-based toxins inert even as it removes them from soil. Always cautious in his statements, Qing also pointed out that cleaning up heavy metals presents a different set of problems from the petroleum pollutants he has studied. So a lot depends on what exactly are the issues facing the soil.

Still, if this proves true in longer-term research, it is an immense breakthrough—it in fact helps answer the most common question I and everyone asks about hemp phytoremediation: "What do you do with the hemp once it removes toxins?" In other words, if hemp cleans up a mine site, an ex-GMO cornfield, or a nuclear spill, isn't that hemp now contaminated? Part of the answer resides in mycelial and similar solutions that can break down soil toxins. But if hemp itself has chemical transformative properties, the world eagerly awaits further studies.

Examining the return of the biomaterials era from a higher altitude, I hope we've established the basic point that, from here on out, it's imperative for every product of every kind in the worldwide industrial pipeline to be completely regenerative and healthy to everyone producing and enjoying it. This includes the farmers and customers of course, and also the packagers, the delivery drivers, and even, one supposes, the seed drill renters. If our paradigm-shifting intentions turn into manifestation, sooner rather than later, I think we fun-loving and sometimes lovable *Homo sapiens* might have a shot.

Whatever fiber apps wind up proving the smartest long-term play for a regional cooperative of independent farmers, it looks like we can't get around the fact that we need at least 3,100 acres to feed a fiber-processing system. The cats will have to at least communicate, if not be herded. The technical learning curve involved in getting at the hurd, at least, is not insurmountable. But, as we're about to see, making a co-op work in a way that keeps its members sane enough to continue working together can be.

CHAPTER FIFTEEN

Herding Rugged Individualists

Fat Pig Society Workers Co-Op,
Fort Collins, Colorado, 2019

*In affairs of this description . . . the first essential is to
study the psychology of the individual.*

—P. G. WODEHOUSE, *Very Good, Jeeves*

I f we're going to herd farmers toward forming regional cooperatives to handle the fiber side of their harvest, it'd probably be
valuable to recount what makes an actual working co-op tick. To
do that, we'll report on a day in the life of the first modern organic hemp
cooperative—the Fat Pig Society workers co-op—during my most
recent pilgrimage to the co-op's Fort Collins farms and greenhouses.
Even though the FPS is a cannabinoid-focused co-op, the lessons are
the same. And even if many of our putative fiber co-op's members aren't
doing much more than opening their gates for the fiber truck once per
season, a portion of any co-op's members is going to have to deal with
the organization's operations. When you look at successful co-ops, you
quickly realize that the more engaged co-op members are, the better.

The first modern cooperatives arose in Scotland and England in the
18th century. They emerged from the transition between serfdom and
modern agriculture. That was 300 years ago. In my research, I've been
shocked to learn how deeply co-ops are integrated into today's economy.
One hundred thirty million Americans belong to at least one co-op, and

25 percent of American electricity is generated by either municipalities or electric cooperatives.[1] Worldwide, there are three million cooperatives involving 12 percent of humanity.[2] I get 90 percent of the food that I don't raise myself at the all-organic Silver City Food Co-op in New Mexico, founded 1974.

Some agriculturally focused co-ops are massive. Blue Diamond Growers, launched in 1911, had $1.67 billion in almond sales in 2016. Organic Valley, producer of eggs and dairy products, has 2,000 independent farming members and 500 employees. The point is not just that cooperatives can be hugely lucrative. It's that the difference between these billion-dollar entities and publicly traded billion-dollar companies is that a cooperative is not owned by shareholders or hedge funds. The revenue stays with its members.

There are so many categories of co-ops. I can't tell you exactly how many, because some co-ops are an amalgam of several different ones woven together. The Fat Pig Society is a worker co-op planning on branching out to include a producer co-op. The rules can vary widely among different types of co-ops. This, unsurprisingly, has led to lawyers who specialize just in co-op nuances, and it can get complex: One attorney suggested to a group I've been trying to corral into a co-op that it form "in part [as] an agricultural marketing cooperative and in part a worker cooperative . . . this would integrate workers into the ownership and governance of the cooperative, avoid servant laborer status for the workers, [and save] a lot of money in tax and labor law."

That alone is a full meal of information to digest. If you and some fellow farmers are considering a fiber (or other) co-op, you've got your background research and legal due diligence to do.

A co-op, of course, is not the only righteous way to go—we're mainly discussing the model here as a way for independent farmers to make something with their tons of fiber. We've also talked a bit about B corporations and profit-sharing LLC models. In Vermont, we started our enterprise as an LLC with a mission statement that flat-out said we intended to evolve into a cooperative model, once we realized it would take beyond planting season even to land on which type or types of co-op we wanted to be:

> Our goal is to cooperatively produce healthy products (for
> our own families and communities, and for more than just

humans) that benefit the local economy and soil (including the farming economy) by promoting a regional and sustainable industrial loop. Therein we hope to provide a regenerative planet-wide economic model that might help heal soil, mitigate climate change and make room for healthy, affluent rural communities to lead the world toward long-term peace.

Lovely, idealistic stuff. But in practice, a co-op is a business. Just a different kind of business. Like any venture, a hemp co-op likely faces years of challenges just to keep itself functioning. I don't really care if an enterprise is called a co-op or something else. If it operates by Bill Althouse's basic goal, I'm a fan: "It'd be nice for farmers not to get screwed for once," he told me as the FPS was filing its formation paperwork, back in 2014.

The other essential thing to know about the modern cooperative movement is that it operates according to these seven principles, sometimes called the Rochdale Principles, for the British co-op that first applied them in 1844. Any co-op member anywhere in the world will know about these principles. At least some of them. Like the Ten Commandments, most folks will probably be able to name about five of 'em.

1. Voluntary and open membership.
2. Democratic member control.
3. Member economic participation. [This and number 4 are the keys that keep revenues among members.]
4. Autonomy and independence.
5. Education, training, and information.
6. Cooperation among cooperatives. [This principle explains how the FPS markets its main cannabinoid product, called Free Hemp. It's not exactly free, but it's free of branding, corporate investment, and marketing other than word of mouth. It's either sold directly by the FPS, by fans of the product who buy a case or three at a time, or at other co-ops.]
7. Concern for community.

The goal of these principles, as Wendell Berry knew from his family's formative role in Kentucky's tobacco co-op movement, was power in numbers—for cobblers, chip makers, hemp farmers, makers of anything. Screw the middleman, rather than the farmer.

———————

I was overdue for a Fat Pig Society field visit in March of 2019. I'd been following the co-op's on- and off-field efforts since before its formation, and with processing done in Vermont and my return flight to New Mexico stopping in Denver, the stars aligned for my first field visit in a year. This time, I found myself hurrying nervously up the Colorado Front Range because of a troubling text I'd received from Iginia Boccalandro upon landing in the Rockies. Before leaving Vermont the previous day, I had messaged Iginia that I might be running a day or two late following the blizzard in New England. She replied, "Please get here tonight, not tomorrow. It's Bill's birthday. We need your energy here."

That should have been a red flag. But it was one of the nicest things anyone's ever said to me. Plus, how bad could things be? Still, I probably wouldn't have harvested a crop or created a product without the selfless guidance of people like Iginia and Bill. Needless to say, I drove up to the FPS farms and greenhouses in Fort Collins, 40 miles from the Wyoming line, immediately upon landing in Denver at sunset.

Up until that text, I thought I'd be reporting on good news about the co-op. I had reason to be bullish. After four years of the usual entrepreneurial struggles, the FPS had apparently beaten the odds and gotten on its financial feet. The last time I'd seen Iginia, at an industry conference a few months earlier, I'd awkwardly offered to pay for her breakfast. I knew that for three years the co-op had been treading water, its members unpaid and busted flat. Why would season four be different? So I was more than a little shocked when she told me, "It's cool, I got it. We're all getting paid two grand a month plus our food and housing costs. We've got fifty grand in the bank."

I considered Bill, Iginia, and the team to be integrity role models. That it was actually working out fiscally frankly blew my mind. When I had woken the FPS farmhouse with marginal Coltrane four years earlier, the co-op's then four members were sharing the best clothing so one person could be presentable in town, at expos, and at meetings with bankers and potential members.

A few years later, Iginia told me over that conference breakfast, one white-label crude contract alone was generating $120,000 annually. Best of all, contract farmers, some of them on track to become members, were already getting paid.

The co-op, in fact, had paid farmer Will McDonough, 44, of Wimo Farms in Berthoud, Colorado, $75,000 since my last visit, plus provided him with $75,000 in clones for his next planting. McDonough, a navy sub vet and in civilian life an organic raw dairy farmer and employee of the Department of Veterans Affairs, told me, "Farming hemp has elevated my family's income level substantially and saved our retirement. I can call it a wonderful relationship because I don't see any end in sight."

When he met Bill and the rest of the co-op team, McDonough said, "I was already passionate about building the small-farmer economy. And we shared organic principles: I fertilized the fields from our own cow manure. It was a good match."

Had the Fat Pig Society cooperative achieved only this one goal, I'd still call this story a celebration. Farmers were making a good living. This was the unwavering aim of the co-op's founders since before hemp was federally legalized. I can attest to this because I knew Bill and Iginia prior to the Fat Pig Society's formation, back when we all lived in New Mexico in the 2010s. Iginia was a Rolfer and ran a speaker program called the Carbon Economy Series, which is how we met. We depleted a lot of the strawberries that Bill grew for his booth at the Santa Fe Farmers Market (the berry vines draped down according to a trellis style of his design), while discussing their plan to keep farmers from getting screwed for once via hemp. A few years later, their real-world organic hemp cooperative was already churning out top-shelf product and paying farmers top dollar.

Hooray, I thought. Right as I was finishing up this book, I had a text-book case study for the persistence and patience preached in its pages: a four-year, small-farm journey to liquidity. In the realest sense, the FPS members had been conducting three years of R and D. They'd learned lessons of cultivation, processing, even packaging.

Four years is a long time to stick anything out. Twenty-two percent of small enterprises fail within a year (not the more commonly reported 50 percent, but still a lot).[3] My mistake, it turns out, was assuming financial stability meant organizational stability. Money might not buy happiness, but, I thought, doesn't it solve all business problems?

I wanted this to be true, especially for a co-op: This was the model I yearned to see replicated in hemp-growing communities all over the world. Indeed, I'd been trying for several years to launch one in multiple states. I'd forgotten that the hardest part of this enterprise would be the human side; that cultivating people would prove more difficult than cultivating and marketing top-shelf hemp, forming a co-op, or becoming fiscally buoyant.

Optimism was my dominant state of mind as I drove north toward the plump porcine logo on the FPS entrance sign. Some unnamed social crisis notwithstanding, I missed my mentors and I wanted to see how, if at all, the cooperative had migrated in practice, principles, or spirit as it was apparently on the rise. When she'd apprised me of the promising financial developments at that conference breakfast, Iginia had also told me that the co-op was on the verge of expanding, almost on schedule, to its phase-two plan of adding a producer co-op to the existing worker co-op.

"We want the worker co-op to do farming efficiency, breeding, product development," she'd told me. "Basically to be a think tank. The goal being enrichment of the producer co-op, composed of hemp farmers."

As a result of this remarkable recent progress, I thought I'd be asking questions about marketing, tax issues, and growth plans for membership and acreage on this visit. I wanted to be bored by the mundanity of success. Instead I learned that the success of a co-op is much less about its operations than its personalities.

Loving these people like kin, news of FPS unrest, though lurking below my optimism, was jolting. In fact, it's still a bit painful to recount the turmoil I witnessed over the next 2 days. I mention this for two reasons. First, because the FPS, six months later as this book is going to production, lives to fight another day, which means the way its members handled a core crisis provides a model for how other co-ops might address social engineering issues. Second, because I believe this co-op's members deserve some kind of integrity and transparency award. On this most recent visit, they were aware I was researching a book chapter about the functioning of a real-world hemp co-op. They knew I was going to report what I saw.

When she and Bill decided to form a co-op, Iginia was aware that managing people would be the hardest job. In fact, it *was* her job, from the start. The daughter of South American engineers came in with some essential perspective: the awareness that the world's most successful co-ops have identified the people problem and learned to address it as a top priority.

"The Mondregon Cooperatives of the Basque region [of Spain] have recognized that it is a career track position to be the social coordinator," she told me four years ago. "This is the person who checks in with every member in her section, every day: every manager, every intern, every engineer, every custodian. These guys have twenty-eight billion dollars in assets, and still there's a social coordinator for every eight people. So we decided to replicate that. As you know I'm that person." Indeed, I did. Iginia was the Chief Cat Herder. But the FPS's big problems arose four years later when Iginia realized she was also one of the cats.

Iginia met me at the camper as usual as I pulled in when the moon was high in the sky that March evening. What struck me as unusual was that, despite the relatively early hour, she took my elbow and walked me away from the farmhouse. We headed straight to the main FPS greenhouse for a private chat before I even encountered the rest of the crew. As we pulled back two greenhouse flaps, gaining 30 degrees within three steps, she said, "I knew it was going to be this way. I just didn't realize how bad it could get, so quickly."

"How bad what could get?" The moisture in the greenhouse air was itself a second wall through which we walked like wizards at a train station.

"The people problem," she said, sitting us down at a picnic table surrounded by plants of all sizes.

"You mean Bill," I said, wiping my brow.

"I mean Bill, myself, Yamie, Gavin, and all the volunteers," she said. I noted that she had listed nearly everyone who was or ever had been a member of the co-op. But then, after a significant pause, she added. "Mostly it's about Bill and me, having crossed a line where we're having difficulty working together."

Then she took a breath.

"That social coordinator's job [in the Mondregon model] is to ask personal questions, like 'What's really going on with you?' and 'Who's pissing you off?' As well as business questions like, 'What could be done better on your shift?' and 'Do you know about the big co-op board vote coming up?' There have been studies about this: It's an essential co-op survival tool."

They must be doing something right over in Basque Country. The 63-year-old Mondregon Co-op group has 74,000 members working at 257 companies. They do everything from making bicycles to offering engineering training. Equally important, 70 percent of the co-op's members vote in co-op elections. That's 15 percent higher than voting rates in US presidential elections.

"The model has been replicated all over Latin America," Iginia continued. "That's where I learned about it. It's all about the team feeling empowered."

It seemed so sensible, so natural. I immediately started trying to work out where else the social director appears in nature. Is there a gorilla Iginia? Possibly. And cattle folks tell me that there's a sort of auntie for the herd, making sure someone's on lookout, and all the calves accounted for. Regardless, I had little doubt she was right about the importance of the position. I don't speak lightly in my advice on this front these days, because I've felt the pain of crossed wires and mixed expectations within an entrepreneurial group. Trust is one thing, and essential, but it's just one building block for a successful working group. Two or three people might have fairly different opinions about what amounts to a full workweek, for instance. Or the big one: How long you all might have to work before getting well paid. No one I've met is immune.

When things have gone well on any project of which I've been part, whether as entrepreneur, consultant, or researcher, it's because every member of the team was jamming in sync. Everyone stepped up without having to be asked, knowing the others would often do the same. When the project hit bumps, it was usually because there were communication issues.

Okay, I got it: A co-op is wise to have a social director position right from launch. Non-discussed expectations, man, what an enterprise killer. But I didn't fully grok what the new issue was for the FPS. When there's a difficult, challenging personality, and everyone recognized that Bill can be a difficult personality, the social issues might come to a head sooner. But they always do. That's because everyone's difficult. It's why I live 40 miles from the nearest town, surrounded by attack goats.

"I get all that," I said to Iginia at the picnic table. "Why are you telling me this now?"

She sighed. "What we realized a few weeks ago is that we need to have a member who is Bill's keeper," Iginia said. "Someone other than me. We know each other too well for me to be his social director. Or my own."

Now I understood: The co-op required another social director, maybe two. A member whose job is just to deal with the co-op's chief farmer and idea man. And probably one to assist Iginia too. Social director was not her only position. She had a half dozen other roles in the young co-op, including membership and volunteer-outreach coordinator, and product formulator. She wasn't a bad farmer, either.

Bill and Iginia have known each other for a quarter century. Imagine your social engineering task, if you're the social director for a newly formed co-op comprising geographically scattered farmers feeding a fiber-processing facility.

"Bill is a genius," Yamie Lucero, the newly inducted FPS member, had once told me. "He can design and execute on a large scale. It's just not a personality that is geared to consensus thinking. Which is what a cooperative is."

Right. Cooperation. We start learning to do it in kindergarten. It is, you could say, the core curriculum. And democracy is a core principle of all co-ops. But oh, man, the G-word. Whether one declares Bill a true genius or a sensitive artist whose PTSD bubbles forth every now and then (Iginia characterizes him as a genius, too, as do I), the situation seemed manageable to me.

"So bring on a new member," I said. "Is that such a big deal?"

For a long moment Iginia said nothing. She just looked at me, and her eyes welled. I was floored by her next words.

"We're discussing dissolution," she said. I felt as though I had been punched. "The situation had gotten so bad that we brought in a co-op-focused consultant just the other day. It kind of went well, and kind of not. That's why I rushed you up here. It's a critical moment."

As social director, Iginia was doing her job. Especially because she was one of the members having a critical social issue, she recognized the need for a grown-up in the room for a few days, watching and asking questions. Someone other than herself.

"Much as I'd love to be the benevolent matriarchal dictator, that's not how it works," she said, and erupted in her massive, contagious laugh that sent the nearest plants dancing. The consultant's conclusions, in Iginia's words, "forced Bill to acknowledge he can be difficult, and forced me to acknowledge that we wouldn't be here without him."

The takeaway here is a reminder that every relationship requires constant upkeep. Heck, everyone's always evolving as individuals. I'm only marginally the same guy who started writing this book. Even as we accept that humans can be hard to get along with, especially in a business environment, I believe farmers deserve some slack. That's because their work is so important to us all. Sequestering sufficient carbon in increasingly healthy soils is the difference between humanity's surviving and not. No pressure.

And it's not as though Bill is a monster to be around 95 percent of the time. Even Iginia says, "His soul is a flower. It's appropriate that he developed flowers and sweet fruit for years. That's who he is, at core."

He's also teaching a new generation an old way to do business. Work ethic is paramount. Don't even try to get a FPS member to go out for some live music during harvest.

"What can I do?" I asked there in the greenhouse as we wrapped our private debriefing that evening.

"Be here," she said. "Do your thing."

She was referring to the fact that I have a Get Out of Bill's Wrath Free pass for some reason, and I often use it (like when I play sax at dawn). Yamie and Iginia can't believe the things I call him on—we recently disagreed on a plant patent issue, for instance. They quietly congratulate me afterwards as though I have just gone the distance with the Champ. This visit, Iginia wanted to me to watch, and when necessary, speak up.

The next morning we sat down with the complete FPS crew (now down to three members, from five) in the same greenhouse, indeed at the same long white trichome-covered table. This is where the workdays were based at this point in the spring: here and in the ethanol-processing garage. In more ways than one, I spent most of that 2-day visit really getting at the root of things with the core FPS team. I did that by working beside them on their crop. Like any start-up, they had no time for anything but an actual day in the life of the FPS. There was no lipstick for this Fat Pig.

Alongside a rotating string of volunteers of all ages working for product (a four-ounce jar of Free Hemp with 1,000 milligrams of organic, farm-to-bottle CBD infused in coconut oil, costs $80), I and the FPS team planted keiki clones just trimmed from bushy mother plants. Bill's technique for ensuring that the keikis took root was different from Iginia's, but both styles worked well, and they communicated their suggestions to newbies like me respectfully.

When that task was finished after a couple of hours, we relocated more mature plants to their final, larger pots. These would be their homes until they landed in someone's transplanting machine or were hand-planted. There's nothing like hours-long stretches doing satisfying if repetitive work in an oxygen-rich, sauna-moist, plant-energy-dominated environment to get folks really talking. Yamie and Iginia quickly got on to the deep topics. Bill's emotions, as always, were conveyed in code.

Grief, too, played a role in the outpouring. Since my last visit, the co-op had lost the architect of its anti-marketing strategy and my good friend John Long to cancer. A fifth early member, 28-year-old Gavin Lim, had split, needing a BFB (Break from Bill). His membership was bought out by the remaining members, though he returns to help out periodically.

When the work allowed, we really talked. Hemp clone work and ethanol processing permit fairly regular bursts of chatting, stretching, and jamming between bouts of concentration. My friends filled me in on bottom-line and social issues while we restocked flats of clones beside the transplanting tables, or while a few of us dashed off to the garage when the timer sounded to indicate it was time for the next stage in the ethanol processing cycle (stirring the flower material, for instance).

At first, I was surprised to learn that most of the FPS's current challenges were the same ones that every plan B hemp enterprise was facing. It could fairly be described as routine turmoil. The classic, harrowing life of an entrepreneurial venture. I'd seen it before. Indeed, I was living it.

They'd made it this far. If it hadn't been for Iginia's heads-up the previous night, I wouldn't have been worried.

The co-op's new bankers, for instance, visited the FPS farm on my second afternoon, meaning the ranch house had to be cleaned up. In the plant realm, just six weeks earlier, a sudden infestation of spider mites had decimated 86 of the co-op's 90 prized mother plants, right when they needed to fill thousands of clone orders.

There was the usual debate among the co-op's members about how to best price their core value-added product, Free Hemp. And, of course, there was the normal suspected rip-off of co-op genetics by a former collaborator: Bill's near fisticuffs with the dude was averted by last-minute Iginia diplomacy. All par for the course so far. Tractor-maintenance issues also reared their noisy heads. Just because you're a righteous, organic co-op doesn't mean FLOATER goes away. It was spring. The co-op's outdoor fields had to be prepared.

Most interesting to me was that I had arrived at a good time for learning about alcohol processing—literally the inner workings of the cold ethanol model. These were the folks who taught me how to decarb. So I figured they'd be promising candidates to demonstrate ideal models of reverse moonshining, which is how I think of ethanol extraction.

On the second morning of the visit, Bill met with a retired Colorado State University materials engineer, now an FPS neighbor, to see if they could rethink ethanol extraction. Specifically, Bill was seeking a finer mesh netting in which to contain the co-op's hemp flower harvest within the machinery's hopper. One that could be more easily removed and cleaned between processing runs.

"Most ethanol setups have a wide cage inside that gets immediately clogged with flower material and is hard to get at," he explained to me while I looked, slack-jawed, at the CAD designs he and the engineer-neighbor were poring over. "I want to see if we can find the right gauge bags to hold our [hemp] flower, which we can remove and wash every processing cycle." Just Bill revolutionizing another industry. Again, nothing new here.

It wasn't until lunchtime that I witnessed any evidence of the current crisis, an eruption between Iginia and Bill surrounding the consultant's conclusions. Bill thought, as many of us would, that the outlay of nonfarm expenses for a conclusion that questioned his behavior was a waste of money. Channeling her study of the Mondregon model, Iginia stood her ground, maintaining that a new cat herder had been called for. I slipped away to let them converse at high volume while I played saxophone in the greenhouse.

Later that evening, we sliced Bill's birthday cake in the farmhouse kitchen where, even during lean times, the FPS team had fed me and countless volunteers hundreds of meals, as though the pigs were already fat. If things weren't overly festive, they had calmed down once the sun had set. I like to think that the appearance of the alto horn during the afternoon had helped. I asked the co-op's brain trust if the mission was as clear as it had been when they launched the co-op in August 2014.

On this there was no debate. Bill's mantra never changes. "You know the answer to that," he said, adding a scoop of à la mode to his slice. "We've achieved our goals when every organic hemp farmer in the state of Colorado is receiving 100 percent of the retail dollar from their crop after expenses."

"That's why we started paying farmers from first dollars in," Iginia said. "That's why we exist."

That's also remarkably close to plan A for this enterprise. Will the Fat Pig Society stay together long enough to achieve its goals? Even its members are divided on that question. As I pulled away at the end of that visit, Iginia again left me with the impression that she was near her limit; that the FPS experiment would be on the brink of ending were it not for everyone's commitment to its mission.

"Shared clothing we could face," she said. "Now we'll see if we can resolve the human issues."

Yamie, by contrast, believed the FPS was "over the hump" and ready for its planned next stages.

"I think we're hearing one another," he told me as we transplanted the final plants before I bugged out, a little shaken, and headed home to my family.

For his part, during our interactions, Bill spoke of nothing but plants, co-op finances, and processing engineering. That is, except for one moment, when he was giving me a maintenance lesson out by the ethanol refrigeration unit, which needed some kind of adjustment.

He said, "Man, this wasn't supposed to eat up half my day." And I said, "Good thing you've got such a strong core team here—Yamie can handle the processing run and the volunteers in the greenhouse while Iginia entertains the bankers."

Bill set down the rag he was using to work a tight valve, looked at me over his slipping glasses, smiled slightly, and said, "Yeah we've been at this a long time." That's Bill-speak for "I love these people and don't know what I would do without 'em."

I remain worried about the Fat Pig Society and its members, but overall confident in the entity's endurance. For one thing, the core members are united on another key point beyond the core mission. And that is firm belief that the mission is best executed under the cooperative model. For us co-op idealists, the message might be to enter with our eyes open. It could be after we clear a few hurdles that things get real.

———

If the Fat Pig Society is an accurate model, a co-op achieves success through a combination of relentless work and good karma. The FPS bread and butter is a $10,000-per-month (and growing) crude order from a pet food CBD outfit called Suzie's CBD Treats, whose product FPS helped formulate.

"This saved us," Iginia told me as we ate Bill's birthday cake, our fingers still fragrant and sticky from the day's work even after showers.

As with other early FPS angels, Suzie's CBD Treats' founder Richard Squire, who had previously founded and retired from the successful Breckenridge Brewery, saw the importance of the farmer-enriching FPS mission. Most people do once they spend a half hour on the co-op's 10-acre farm parcel just outside downtown Fort Collins.

"We helped Richard with some health issues," Iginia said, which included a since-shed opiate prescription resulting from a back injury. "And then he came out of retirement to work with us. It was win-win. When you're not obsessed with immediate profit, you can operate largely on belief in your mission."

This is not the only time that karma has buoyed the FPS. Back in 2016, a jar of Free Hemp improved the quality of life of a couple's son—the husband was a food safety consultant and the wife was a food safety lab scientist. Together they helped the FPS design its commercial kitchen operating procedure.

"That was ten thousand dollars' worth of consulting, gratis," Iginia said. "In Venezuela we have a saying: 'One hand washes the other, and together they wash the face.' It's cooperation like that that keeps a fledgling co-op alive."

And it all came from appreciation, not hostile takeovers. You can read the relief spreading across Iginia's face at moments like this. This is a woman who considers bobsledders coddled wimps—not a fragile gal. But even as she worries if the co-op can transcend its current personnel crisis, she relishes the achievements on the entrepreneurial side. We both thought revenue would be the hard part for the FPS. Let's see that taught in biz school: karma investing.

As a co-op, the FPS can't accept investment in exchange for shares of the enterprise, but it can accept in-kind services like the commercial kitchen consultation, as well as loans and friendly real estate exchanges. The result of good work combined with this kind of angel support is that phase two of the FPS—launching the producer cooperative with the aim of including more farmer-members—is slated for mid-2020.[4] I watched Iginia in her office preparing to post the required notices for the expansion, per Colorado co-op rules. This was a good sign. Moving forward with plans despite some internal issues.

"My work lately is all about increasing efficiency," Bill had told me as we were rooting keikis. "How do we harvest and process one hundred pounds of flower in five hours? What do we charge for that?"

He's thinking ahead. If the FPS is ready with a game plan and membership increases as a result, so does the bottom line for farmers.

To demonstrate how this is so, Bill ran the numbers for me. "The farmer can now process four thousand pounds wet, per acre, with our genetics. That's two tons, which in drying and processing goes down to half a ton, and then again to ten percent of that for a crude of fifty pounds."

The next bit of math mandates deciding how much of that crude goes to wholesale and white labeling, and how much into Free Hemp and the rest of the expanding FPS value-added product line. CBD crude, as of this writing, is worth about $2,500 per kilo, or $1,130 per pound. Producing those 50 pounds of crude per acre means a 2019 Fat

Pig Society farmer's crop is potentially worth, before any value-added product, $56,500 per acre. That's assuming current wholesale prices hold, which as we've been screaming is a big assumption, though less so for organic product.

Not too shabby. Especially at 5 or 20 acres for farmers like the McDonoughs. The clone-selling side of the FPS business is also a reliable source of cash flow. And, Iginia told me, they're sticking with their long-standing anti-marketing policy for Free Hemp; it's only retailed by co-ops.

"And we still prefer to provide folks with a case, to distribute to their friends and those in need of high levels of bioavailable CBD," Iginia told me. I refer high-CBD requests to the FPS all the time.

As Suzie's CBD Treats' Squire can attest, lives have been improved by the FPS harvest already. That's not a bad epitaph for a person or enterprise: IMPROVED LIVES. For the long-term, though, none of this matters unless the core social issues are resolved. I'm hoping the FPS members' grit, combined with the support of angels and cheerleaders, can see the co-op through these hard times. I'm far from the only one who recognizes the stakes are too high for the enterprise to quit on the cusp of success. The FPS is a role model for the farmer-first renaissance. And boy, when it comes to both production and especially marketing, the members sure are following Dolly's advice to do things "different."

Different for now. I can't wait for the Fat Pig Society model to become mainstream. When it does, when a group of farmers and processors is willing to think beyond current quarter profits and into long-term prosperity, I bet it will compete with and even outperform the old Big Ag model. Maybe we really can have Main Streets again.

Have a Compelling Backstory

Before you leap into the hemp party, there's one more reality to keep in mind. The effort to establish independent regenerative hemp farming as the industry standard overlaps fortuitously with market demand across the consumer landscape for non-mass-produced, non-ultra-pasteurized, non-chemically soaked products. That means millions of customers are looking for a reason to buy a craft product. More and more shoppers are becoming label readers. They're on to the junk options.

The very word *hemp* carries with it, in the words of our favorite researchers Ernest Small and David Marcus, "products associated with environmentally friendly, sustainable production."[5] In other words, hemp is already a healthy brand. This is priceless free marketing for regenerative entrepreneurs. You want your brand to enter the arena with a positive reputation. You don't want to work uphill to fix a problem. As you would if you were hawking, say, asbestos. Here we see justification for what otherwise might seem the hopelessly idealistic attempt to take a crack at solving humanity's biggest problems while making a living. Once you realize that hemp is not just your product but a lifestyle brand, you'll be off and running in the biomaterials renaissance.

Whatever your story, tell it right on your bottle or accompanying literature. "I cultivate hemp because . . ." Hopefully, your early-adopting base will digest the story behind their purchase rather than focus solely on the price tag. This enterprise backstory is what Michael Pollan calls "Supermarket

Pastoral." You know, the way you think the chickens who laid your free-range eggs are getting massages and profit sharing.

Your Supermarket Pastoral might be a strange cannabinoid ratio you stumbled upon in the feral ditch weed that has grown in your Nebraska yard for generations. It might be infusing your product with your grandmother's favorite essential oil. Maybe it's the low number of carbon miles[6] embedded in your product, or the large number of farming families your enterprise supports in style.

Better still, brand the fact that you *are* the farmer. A key part of the game plan involves educating buyers about how important regenerative farming is to everyone's well-being in this new abnormal of climate chaos. Educating customers is key, because our entry prices might have to be a bit higher than bottles of McCBD. Economists call this the *first-to-market principle*. I call it the non-poison surcharge. The hope is that it goes away when regenerative becomes the norm.

My own Supermarket Pastoral messaging approach is including the phrase "harvested without petroleum"[7] on the product label. Plus, I do this messaging on plant-based labels backed with the aforementioned nontoxic stickum. And I'm paying for 'em, let me tell you—nearly 50 cents per label at this stage of the game. Worth it, though, for Earth and marketing reasons. It's generally a clever move to get creative with (and then shout about) the beneficial practices embedded in your very modus operandi. In fact, the best part about regenerative practices (after saving humanity, I mean) is that they feed your bottom line.

If your Supermarket Pastoral elevator pitch is genuine and you execute it effectively enough to make a sale, you're likely to make it again. Once they experience how awesome your cereal, tincture, hair conditioner, or dog treat is, your now-loyal customers will see that actually they are not paying more. They are paying for something different. And better. Suddenly they find they're not even mere customers. With their purchase, they're investing in the species' future. With their awareness that, as Wild Bill says, farming is going to save us, they've become an active part of the regenerative economic pipeline. They are building their own communities.

CHAPTER SIXTEEN

Green Chile Hemp

Find out who you are and do it on purpose.

—DOLLY PARTON

Every US state has an official bird, flower, and song. Only New Mexico has a state question—"Red or green?"—which refers to the flavor of chile you'd like on your chalupa or burrito. We don't take this question lightly. You really do get asked in Land of Enchantment restaurants. I'm a green guy, straight up.[1] It's our built-in brand.

In his hemp venture, my colleague Lew Seebinger was on to this immediately. He launched red and green product lines in Albuquerque back in 2013. Some Seebinger Hemp products have famous Hatch green chile in them. High in vitamin C, chile has long been used in local poultices and tinctures. In our region's *curanderismo* (folk healing) tradition, it is believed that chile helps increase circulation, and that its capsaicinoids can be helpful for muscle relaxation or joint pain.[2]

When your field is harvested and your hemp heart or CBD product is processed and packaged, folks have to know about you. Whether you do 10,000-, 100,000-, or 10,000,000-unit product runs, we're going to take a leap and presume you'd like to sell them. And so we delve into the world of branding and marketing. Believe it or not, these can be regenerative too.

From what I've seen in the young industry, I'm convinced that it will help immensely with your value-added product marketing effort if you, like Lew, make your life easy: Find your built-in brand. Your most effective elevator pitch allows your product to sell itself because it's so clear, so obvious. And in the regenerative value-added space, it's part of you and your place. It's not just about having an effective Supermarket Pastoral backstory: It's demonstrating that you live it. Your bottle should explain in an instant why you and your surfing buddies, let's say, created a hemp-based sunscreen product, and why it's your first choice to slather on your nose before heading out to catch a wave.

It might be your family story: "Going Legit Hemp: our first legal product after three generations of research." Might be your personal mission: "Saving the Family Farm" brand tincture. Maybe it's your Everyman bottom-line reality: "Tuition Hemp Hearts: Puttin' three kids through college while fighting climate change."

I figure I've heard the charming backstory to nearly every legitimate independent hemp product that's crossed my path. The strongest enterprises all have one. My own, obviously, is a small batch, farm-to-table feel. Tonia Farman and Gregg Gnecco at Hemp Northwest are motivated by a desire to rejuvenate a cucumber farming economy that left their Pacific coast region for Asia a generation ago. Anything personal has appeal. But you probably couldn't live your brand more than Janel Ralph, founder of Palmetto Harmony in Conway, South Carolina. Her CBD product line's Supermarket Pastoral is essentially, "This product saved my daughter's life."

Now, Janel, 44, might not describe her daughter's rather remarkable health-improvement story as her brand. But the beautiful 10-year-old Harmony's name is part of the company name. And everyone who knows the company's products follows Harmony's story. Her condition is called lissencephaly. It is characterized both by low brain activity and seizures, among other symptoms. Most children with Harmony's type 2 condition don't live past age three or four. Harmony is now approaching her 11th birthday.

"When conventional treatment for her disability was doing more harm than good, I discovered that CBD helped," Janel said. "But at the time, in 2014, I couldn't get it for her reliably here in South Carolina. I had to start a company."

My recent visit to Palmetto Harmony headquarters began with Janel showing me Harmony's brain scan, a detailed mapping called a quantitative electroencephalogram (qEEG). She pointed to shaded areas of activity that were inactive before her CBD regimen began.

"Her doctor said this latest scan is beyond what she is supposed to be capable of doing, short of divine intervention," Janel said, waving the laminated sheet and speaking with a slight catch in her throat. "The only catalysts for the improvement were adding CBD and removing ninety percent of the pharmaceuticals."

"Does she seem happier to you?" I asked. Anytime I see Harmony, she seems pretty chill.

Janel, who is a pretty good friend of mine, shot me an *Is the pope Catholic?* look. "When you go from hundreds of seizures per day down to periods of a month with no seizures, no bad days, uh, yes, it clearly improves her quality of life."

Palmetto Harmony now employs 27 people in South Carolina, including a greenhouse manager. The plants looked and smelled great when I strolled among them. They are certified organic. Best of all, this independent farmer-entrepreneur controls her product throughout the value chain. All Palmetto Harmony product is produced on-site in a vertically integrated closed loop. Harvested flowers go from greenhouse to processing room to bottle to customer. And the company owns its genetics, in addition to the usual reasons, for a very personal one: It's what works for Harmony.

On the surface this might give some hope to other parents of children with lissencephaly. But for families desperate for any promising avenue, the wise move is to work with a medical professional, as every human body is distinct. Janel's regimen for Harmony has been conducted with the explicit guidance and support of her doctor. For the producer, the key takeaway is to avoid the dangerous world of medical claims, even if your first two dozen customers have told you your product has saved their life. Just do good work, good media outreach, and let word of mouth kick in. Some entities that have not heeded this advice have garnered FDA attention.

There's a lot to admire in the way Palmetto Harmony operates, but my favorite move the company made was to air the first nationally televised cannabis commercial—which you can find on Vimeo—in 2017. It was brilliantly conceived, focusing on wellness for veterans and seniors playing golf. Lots of American flags. My kind of messaging.

Now, I don't claim to be a marketing expert. Tomas Balogh, an Emerald Triangle ganja farmer-entrepreneur whose work I followed for a year in *Too High to Fail*, is always reminding me that the lesson of Nike is to choose a short word, phrase, or symbol that means nothing to most people, which you transform into your own message. He chose Kama Tree for his sun-grown brand. At the same time, I've been handed hundreds of hemp products in the past five years, and if there's a first impression lesson I've learned, it's this: Make the journey to purchasing as easy as possible for your customer. Don't make 'em work.

Unless you want to be an anti-brand like the Fat Pig Society's Free Hemp, I recommend against pulling a symbolic "The Artist Formerly Known As" gag on your label. Just tell your natural story concisely. Let your label and packaging reflect your entire enterprise. And in your social media, demonstrate that you use your own product(s) daily.

Branding and labeling brainstorming is a fun phase in the hemp calendar—enjoy it. In the middle of the season, you can look over your wavering plants, maybe fire up the barbecue, and find the obvious message embedded in your enterprise. Is there a particularly prominent terpene in your cultivar's bouquet? Its mango scent might inform your logo. My new labels feature a goat yanking off my cowboy hat. It's based on a photo of my very first goat, Natalie Merchant, doing just that. Humor is my salvation, so it's reflected in my enterprise's product label.

You might also keep in mind that your farming and business modes, your brand name and your label—collectively, your enterprise's elevator pitch—also has a second job. Since we are in the process of defining what independent regenerative hemp entrepreneurialism is, your messaging carries extra weight. You are an early member of the team that's helping establish our niche in the marketplace. The craft space. So look sharp and tuck in those pajamas: Whether you know it or not, you're a role model.

Varietals, Bouquets, and Tastings

Whether you're creating your product and its storyline or looking to buy top-shelf hemp, here's a word you might want to keep in mind: *terroir*. This is the culinary marketing concept whereby a place's name can only be used if the cheese (or wine, or hemp) is cultivated in that place.[3] In some cases, terroir is backed by international law. This is why California champagne is called *sparkling wine* even if grown with Champagne grapes.

In a largely anonymous consumer landscape, a terroir stamp can be used to significant advantage as part of the independent entrepreneur's marketing arsenal. In fact, it's already encoded in Vermont's hemp regulations. Recognizing that the word *Vermont* itself is a valuable asset, conveying green mountains, Phish, and a Ben & Jerry's craft mind-set, the state's Agency of Agriculture, Food & Markets included a Vermont Hemp labeling protocol in its draft 2019 hemp regulations. To claim VERMONT HEMP on your label, you have to cultivate in the Green Mountain State. This is why my product looks like a little bottle of maple syrup.

It's not hard to see how Vermont's deciders came up with the idea: The state already has an almost ridiculously huge and varied craft beer economy. There's a brewery for every 8,700 people, collectively crafting 151 pints of beer per adult Vermonter.[4] I've rarely seen a Giguere or a Nohl drink the same brew twice.

So much for an anonymous landscape. Chain stores just puzzle Vermonters. And though Montpelier is the only state capital without a McDonald's, this mind-set extends beyond Vermont. In our working lifetimes, craft beer will be a larger market sector than mass-market beer. With $27 billion in annual sales, the microbrew niche is about to cross 25 percent market share, and it's gaining more than 1 percent per year.[5]

Craft beer is thus a great example of the type of entrepreneurialism hemp growers should emulate. The two keys to the niche's success: Microbrews taste better, and they are regionally produced. They are of a place.

It's not just craft beer that's increasing market share. It's craft everything: Millions of folks are craving authenticity across their entire shopping list. Still, at the moment it's easier for marketing departments to push mass product when every message is the same. You can plug it into toothpaste, entertainment tentpoles, or political candidates. As a result, in an era that urges homogenization, leaving all but the dogged with one and a half options for everything from pizza to telcos, a clever independent entrepreneurial play is gourmet distinctiveness.

One of the edges we farmer-entrepreneurs are trying to maintain is the concept of living products. Meaning not totally sanitized products. Our brand balances digital-age know-how and local soul. Whenever I get to this point, I always think of something that our South Carolina farmer Nat Bradford, scion to one of the oldest watermelon growing families in the United States, told me: "When my great-great-grandparents were breeding their favorite varieties a hundred and fifty years ago, they didn't select based on sweetness spectrometer analysis or focus groups. They sliced up melons and choose the ones that tasted best to them."[6]

Another important chord to strike is demonstrating that your product couldn't be grown anywhere else. Might be the mountains of Mendocino, caves of Arkansas, or the Ashanti Region of Ghana. Its terroir determines its scent, its protein or cannabinoid profile, and your values. Not only that, but in our craft niche, your 2022 harvest vintage will be different from your 2021 harvest vintage. We're entering the era of top-shelf hemp varietals.

Visualize your product in the hemp equivalent of a fine wine or craft beer shop. The kind of place you'd go if you wanted to pick up something special for someone's pre-wedding dinner. Even for everyday wellness, as hemp and cannabis markets expand, a significant portion of customers will choose a regional, small-batch offering, rather than the humdrum CBD offerings in chain stores.

Craft hemp customers are the same folks that choose craft beer, a fine wine, or farmer's market herbs today. There will be hemp tastings following farm tours. Knowledgeable salespeople at fine hemp shops will tell tourists in your region, "Well, the 2022 harvest didn't have all the CBN that the previous seasons did, but the overall cannabinoid and terp balance makes it my favorite harvest by the Carbon Sink Hemp folks. Delicious."

When it comes to hemp-industry market share, craft production might start in the lead and never relinquish it. One way to help ensure

A Tip for Hemp Customers

On trade show floors these days, I like to stroll up to a booth, listen to a CBD pitchwoman's energetic spiel about improving sleep or helping dog arthritis, and then ask, "Do you source your hemp from local farmers?" I'm just asking, but after we chat for 10 minutes, more than one pitchwoman has told me she's going to bring the matter up at company meetings, or even change her focus within the industry.

If you are a hemp customer, it's helpful to us, the farmer-entrepreneurs, if you keep this "know your farmer" ethic in mind as you shop. Anytime a hemp hustler can't tell you how or where the actual crop in his product was grown, or whether the farmer is included in the revenue sharing, that should raise a red flag.

If you hear, "Well, I'm not really on the cultivation side," I suggest looking for another product. No righteous backstory? No buy. Also not good: "We buy our CBD on the open market." You can do that yourself, minus the fancy bottle. I wouldn't, though.

A much better answer is, "I'm the marketing director, but I'm on the farm at least once a week. Look how dirty my fingernails are! We cultivate biodynamically from genetics we've developed, and our nine full-time farmers are all included in profit-sharing. We're a B corp, and in the next three years we aim to evolve into a cooperative. The reason I can tell you about the farmers' role is that I'm married to one of 'em."

When you get that kind of answer, you're almost certainly looking at a superior product.

this is to let folks know that by enjoying what's in your bottle, they are doing good for themselves, the families of farmer-producers, and the planet. Stress that this is explicitly because it's organic, farmer-supporting, soil-building, low in carbon miles, and plant-derived. Right out of the gate, you're validating your brand. What customer wouldn't want to buy organic chocolate-covered hempseeds whose label informs her, "You are sequestering 20 pounds of carbon for every serving of this superfood that you eat"?

Cannabis prohibition was an aberration that drops in our laps this incredible opportunity to relaunch regeneratively, from farm techniques through business practices. The more true you stay to handcrafted, organic principles, the more value your product will retain over time. This is the driving force for producers like Vermont Farmacy and Natural Good Medicines.

But don't think that, because you've executed well in the field and in the labeling software, your work is done. As it was when craft beer launched in 1979, part of our collective task will be to teach our customers about how fun it is to explore taste, scent, and texture differences among hemp products that emerge from the diversity of varietals and locations in which we independent craft farmers are cultivating. Already some dispensaries introduce customers to farmers with "featured farm" displays that include videos of life in that cannabis/hemp field.

For seed-oil products, there will be buttery varieties, pleasantly bitter ones, and extremely oily ones. Connoisseurs will stock hempseed oil cellars the way they today do wine and cheese cellars. Flower offerings will also develop top-shelf taste and distinct entourage effect reputations. This is the Dolly Parton model.

If you're detecting a hoity-toity dynamic in this strategy, it is there, and we'll have to find the comfort zone between accessible and top-shelf. Craft products might indeed be more expensive per unit than a box store offering, at least at first. That's to be expected: Dom Pérignon is going to cost more than Two Buck Chuck. But this doesn't mean pricing out of the range of non-gazillionaires. It means pricing in a way that pays farmers well.

Still, how do you convince someone to spend more per unit, when there are cheaper, faceless alternatives down the street at the pharmacy?

Well, Frank Zappa used to ask every auditioning musician, "What do you do that's phenomenal?" With a similar mind-set, you're wise to feature what you do as a farmer that is spectacular (Dolly's "better") and that no one else does ("different"). Highlight that you nurture your soil as though your customers' lives depend on it. Because they do. Explicitly explain that there is more to the cost of a product than sticker price.

That accomplished, how will you get your product to customers? The age in which we live means that digital sales and other direct marketing approaches are likely to account for a major share of sales for the foreseeable future. Even when gearing one's marketing toward folks consciously shopping for distinctive, terroir-based hemp products, the game-changing reach allowed by virtual marketing will likely be an asset for the independent, vertical enterprise.

"We sell about thirty-six percent of our product via the web and [much of the balance] at retail outlets like the Vermont Hempicurean, and Clover Gift Shop and Apothecary," said Colin of Vermont Farmacy. "And Erin teaches CBD yoga classes which incorporate the products."

Whatever it takes. The independents I see succeeding are the ones on the horn with food co-ops or stables, renting space at farmers markets, figuring out shipping costs, hand-delivering product and invoices to stores, staying on top of regulatory developments, and planning social media giveaways.

While the general decline of brick-and-mortar retailing gets lots of attention, Vermont Farmacy's high retail sales percentage demonstrates that wellness-focused products lend themselves to physical shopping. Personal interactions, selecting from top-shelf, small-batch products, discussing it all with knowledgeable staff: These are some of the reasons I think we'll increasingly see high-end hemp shops popping up in our more thriving downtowns—the Napas, Portlands, and Burlingtons of the world. These stores might include products from all sides of the plant, from tinctures to hats to, let us hope, hemp literature. Or they might focus only on fresh flower. Or hemp clothing.

E. R. Beach has been open for business in Ohio with his Hemptations stores for 25 years—I got a hemp-fiber clipboard from him in 2014 that's still going strong. He believes that, just as we're seeing a healthy food renaissance in the midst of a global obesity epidemic, so folks in the

Amazon age will often want to personally touch, smell, and taste a hemp product before buying. His "exponentially increasing sales" trajectory in the past five years has, he said, "allowed us to open more locations and sell more US products."

Palmetto Harmony opened its first dedicated brick-and-mortar store in Conway, South Carolina, in 2018. Janel said its first year revenue was around $100,000. "It's a great business model for a single-family enterprise," she said. Similarly, my own group sees a retail focus as significant for the medium-term future of Hemp in Hemp. Kim and Dana are absolutely convinced that their Tumbleweeds Health Center is going to be a substantial platform for us. I can't overemphasize how much that helps me sleep well. Trust plays in here. They have a seven-year customer base: That speaks volumes in any industry.

All the aforementioned small-batch distribution platforms have something in common: They make economic sense only if producers are not regulated into homogeneity. We are going to require our own set of friendly regulations for nutritive supplements (and other branches of small-batch hemp) distinct from any pharma-style schematic that larger producers or regulators might have in mind. That means we must have our collective craft niche Supermarket Pastoral ready. Embedded in all independent, regenerative hemp messaging is, "We, the actual farmers leading the multibillion-dollar hemp renaissance, recognize that at the global scale there's a move on to sanitize products, including nutritive supplements. Such homogenization leads to our extinction. Might we find space for this ancient, regenerative craft mode? The good news is our products are better."

Defining Regenerative Distribution Parameters

While we're in the trenches with our multiyear effort to establish our product lines, let's take a glance at the scope of the distribution landscape for our niche once our enterprise succeeds. Even the breadth of our enterprise's ambition is a regenerative decision.

As we prepare to manifest the craft hemp market, we have these fundamental questions to address: Do we want one or two major suppliers of hempseed oil, hemp hurd, or CBD to dominate the market? Or do we want many thriving regional providers, elevating entire communities of farmers and production teams to sustainable affluence? The related practical questions for each enterprise are, how much do you want to produce to make a fine living, and how far and wide do you want to ship it?

Maybe it's because I've been spending so much time around plants, the "fifth isn't the only gear" thought keeps coming back to me. In practical terms, how big is big enough? And when would a well-intentioned regional enterprise that is humming along, engaging dozens of farmers, supplying millions of dollars of products, know when it's time to put on the brakes?

First off, what a wonderful problem to have. Humanity is in a much, much better place when "How do we sustain hemp's growth over seven generations?" is our primary question. Our economic mind-set for the past few centuries, let's be honest, hasn't been superlative in this area. You could argue that ignoring limits has been modern globalization's major mistake. Limits don't fit well into the stock market model.

If we take a lesson from the plants themselves, we notice that there are times of day and phases of the year when a portion of a field might remain relatively static or even contract. Similarly, sometimes we humans flower, sometimes we are dormant. Sometimes we focus on our love, sometimes we play defense, sometimes we throw nearly all our energy into providing for the next generation. How do we take the entrepreneur's relentlessly gung-ho spirit, and imbue it with a regenerative protocol?

No doubt there is more than one route to regenerative production and distribution. A lot of it comes back to supporting as many farmers as possible. As Iginia puts it, "We'd rather see a hundred farmers growing ten acres, each making a solid six-figure living while collaborating, than one farmer growing a thousand acres making ten million dollars."

Iginia's on to something: This is how our society works now. It's Balkanized. In the 1990s, 21 million people tuned in to the same commercial running on *Seinfeld* every Thursday. Today, many of us are fans of niche shows or podcasts drawing, if they're hugely successful, an audience of 5 million. The same 21 million people are still watching something. But the disparate media landscape means fans can now support five streaming, grassroots shows when they used to support one. Similarly, five regional producers might serve a region's hempseed food needs where they had once been one—each making a fine living care of a one-million-strong fan base if not five times that. Even 10,000 customers spending $100 per year on your product makes you a million-dollar-grossing operation.

As for when to know your entity is lucrative enough, that's a decision to be made by each enterprise. If it's regenerative, something other than maximizing short-term profit has to be the motivation. Jeff and Wade Lee of the Haloa Aina company, on Hawaii's Big Island, have revived the Hawaiian sandalwood essential oil industry and in the interest of forest health immediately capped production at an amount (120 kilograms per month) that brings in $3 million annually.

"A tree's life cycle is long," the 61-year-old Jeff Lee told me on a fragrant tour of the 3,000-acre forest. "We'd like to watch how the forest develops for a human generation or two before deciding whether it can handle expansion." Jeff and Wade do not aspire to see their sandalwood in Costco. Three million bucks a year is enough, and Jeff said, "Most of it we put back into the forest—our goal is to rebuild a native ecosystem that was decimated by cattle overgrazing."

As we consciously nudge the economy into a fundamentally regional mind-set, this will itself help us to establish our enterprises' ideal evolutionary pace. Maybe there will be some lively discussion about what constitutes "regional" and "regenerative." I'm okay with that. Does regional mean a 500-mile sale radius, or a 5,000-mile one when delivery modes are carbon-neutral or carbon-sequestering?

As a starting point, I'll suggest that our craft market sector define its upper production level parameters at 15 annual tons of product by prepackaged weight. That's a semi-random amount—maybe I'll look back on that number and laugh at its paltriness. In my own enterprise, it would mean 150,000 units of Hemp in Hemp in three-ounce bottles. Even split among several partners and with profit sharing and expenses, that would leave my family with a more comfortable living than we've ever experienced. And it could be done regeneratively on 20 acres.

To the increased regionalization of our economy, I vote yea. Why buy the far-off corporate toothpaste option when the local one will probably be better, and you'll be sequestering carbon with your purchase? Let's also remember that when we buy fungible, disposable stuff from far away, we have little or no idea about factors like carbon emissions and labor practices.

But I've long wondered how successful entrepreneurs feel about the idea of self-enforced expansion limits. While moderating a panel at a 2016 industry conference, I asked this "Do we want one giant righteous hempseed oil company or soap company, or should there be many small and midsized ones?" question of two of my business heroes, David Bronner (CEO of Dr. Bronner's) and John Roulac (founder of Nutiva and RE Botanicals).

Full disclosure: You can sometimes find me chewing on a Nutiva Organic Hemp & Chia O'Coconut macaroon, and I do my washing up with Dr. Bronner's organic soap (almond). My family goes through a gallon of the stuff per month. In fact, I think the west side of the company's production facility is primarily for our use. Both these fellows run undeniably righteous, organic-only companies, and I profiled both in *Hemp Bound*. Both companies are also rapidly growing and each is selling north of $100 million annually. These guys run enterprises that are the anti–Dawn Liquid, the anti–Mazola Corn Oil. But in the organic hemp space, their brand names are ubiquitous the way Mazola and Dawn once were. Nutiva, in fact, is the largest organic superfood company in the world.[7]

It's a testament to the integrity (and confidence) of both John and David that they placed no restrictions on what I might ask them in front of a live audience. In their answers to the question, they both expressed no qualms about regional competition. Instead of appreciating their forthrightness and lofting a softball follow-up, I also broached packaging: "Great—how soon are your companies going to be shipped entirely in bio-based regenerative materials with your delivery and distribution networks at one hundred percent carbon-neutral?"

They didn't blink. John pointed out that Nutiva has a zero-waste headquarters and warehouse where 770 metric tons of waste gets diverted from landfills each year for reuse or recycling. David's answer was, "We're at one hundred percent postconsumer recycled bottles, and we had to fight for that." Both said their goal is a completely regenerative enterprise. I think they liked the challenge embedded in these questions.

They are, after all, entrepreneurs. Entrepreneurial minds are weird—they don't rest for long.

If any large companies are going to become 100 percent regenerative anytime soon, Dr. Bronner's and Nutiva are obvious front-runners. Point is, this distribution issue is another category we have to address if we're serious about establishing a regenerative economy. The farther you ship, unless you're using a solar-powered plane and electric trucks, the more petroleum or coal you use. Someone, one supposes, will have to stock Costco, while there's still a box store model. And I love knowing I can find Bronner's soap from Hawaii to Prague. But for most producers in the coming era, I think regional is the name of the game.

In the end, it is up to us both as producers and enjoyers of cannabis/hemp products to decide what kind of industry (and wider economy) we'd like to help emerge. I'm buying Green Chile Hemp. Carbon Sink Hemp. Moon Cycle Hemp Salve.

"What do you grow around here that's phenomenal?" I'll ask the hemp shop saleswoman in Angkor Wat, Baton Rouge, or Dakar. "Do you process it by ancient modes? Are you the farmer?" As long as it meets a few benchmarks (organic, local, all good ingredients), I'll try the top-shelf selection. Make a mental review. Buy a few extra bottles (or tubes or shirts, ukuleles or batteries) if it's as phenomenal as advertised.

But most of the time, I'll probably think, *Meh, I'm sticking with Hemp in Hemp.* Because as cannabis cultivation legend Ed Rosenthal points out, the best herb you've ever enjoyed is the crop you grew yourself. I hope you feel that about your product. Everyone should. And it'll come out in your marketing. I suggest being humble but proud of your hard work as a farmer-entrepreneur who's trying to help your species endure.

CHAPTER SEVENTEEN

The Friendly Fungus and the Hairnet

The human body [is] an elaborate vessel optimized for the growth and spread of our microbial inhabitants.

—JUSTIN SONNENBURG, microbiologist,
Stanford University

I magine, for a moment, that it's harvest season, 2025. Our web of regional networks is distributing terrific product and sequestering millions of tons of carbon. Farmers are thriving, your group among them. Last season you shipped 11 tons of hemp in various forms, mostly value-added; this year it's looking like 11.2 tons. If this is the state of things, it means we farmer-entrepreneurs have realized a key off-the-field goal: the establishment of workable regulations for our craft hemp niche.

A half decade in advance of that reality, we're all immersed in our distinct phases of the moment, relentlessly inching forward, hopefully having some fun along the way. But if we're going to reach the promised land of an independent farmer-dominated regenerative hemp industry, I hope we have the sheer business sense to initiate our craft sector's distinct regulatory structure.

The reason we're going to embed farmer-friendly hemp production channels in coming policy is not so that we can make extra profit. It's so that we can be unhindered as we provide top-shelf, bioavailable products. It's so that what we are, at core, is allowed to thrive. We are vintners, not cookie-cutter mass marketers. We're going to make sure we can reliably get folks these superlative and righteous salves, boots, and goat treats.

Just as we discussed earlier regarding genetics access, we're interested in production policy that actively encourages a level playing field for those of us producing those 15 tons of product or less per year. That means policy that equally ensconces our craft market's place alongside the fungible wholesale and pharma sides of the industry. And this rule structure will apply for food-grade and non-food-grade products. It's on the food and nutritive supplement side, of course, where the FDA comes in.

Defining our sector by annual production volume is just one way to categorize ourselves. Maybe the craft sector will be quantified by additional variables. Maybe proof of carbon sequestration and organic status can allow even larger producers to participate under the special guidelines we're going to establish. Regardless, "regenerative craft production" sounds like a great section of the Digital Age Homestead Act or the Green New Deal. I also like "soil farming" incentives.

Fortunately, there's precedent for small-batch production rules. The simplest is called *cottage food law*. It's what allows for craft fairs, some catering businesses, and farmers markets—Vermont lieutenant governor David Zuckerman, a permitted hemp farmer, sells his smokable hemp flower at the Burlington farmers market. The cottage rules vary per state. Missouri caps income at $50,000 while New Mexico has no income cap, but a more specific roster of products that qualify.[1] Very few states allow cottage-level rules at the retail level without the additional requirements that a larger food company faces, although some retailers, especially in rural communities, practice civil disobedience in this area—direct-from-farm (that is, under the table) eggs, raw cheese, that sort of thing.

As it stands today, once you get to professional production levels, you start dealing with food-handling and food-testing laws. These kinds of basic safeguards make sense. We all want the products we buy, especially food-grade products and products that touch our skin, to be safe. It's the homogenization of any variance, the "death to all microbes, good and bad" direction of globalized food law for which we provide a countercurrent. We don't provide sterile products. We provide living products.

With hemp, we have the opportunity and obligation to codify our living foods; in essence, to expand the essentially amateur limitations in existing cottage food rules to our 15-ton professional levels (or wherever we land on ceilings). And we'd better do it before the mass market and pharma aspirants decide for us. We've got our work cut out for us. Our friend Roger Gussiaas of Healthy Oilseeds in North Dakota is a perfect example of someone doing food-grade hemp production by the book. That's because at his volume levels (much higher than 15 tons per year), he's got his eye on proposed global food protocols. To say microbe testing eats up a lot of his time is an understatement.

"It's never-ending," he told me as we toured his vast oil-pressing facilities in hairnets and smocks in 2018. Some of his grain storage bags were two stories high. "We spent two thousand hours and twenty thousand dollars last year going through something called the Primus part of the Global Food Safety Initiative. We did it because we feel a retailer is going to ask for these certifications. I'm lucky that my wife and my sister are very good at keeping track of it all. It's more than one full-time job."

I asked if the whole rigmarole made his pressed hempseed oil safer. Taking a look at a temperature gauge on one of his presses, he thought about it for a few long seconds.

"Some of it does," he said. "It involves testing for salmonella, coliforms, those kind of things. But in truth, we have a microbe-kill step when we mildly heat our press to facilitate flow during the pressing process. Regardless, we go through this constant stream of documentation and inspections."

Indeed, on the day of my visit Roger and his team were prepping for yet another safety inspection. Even the calibration of his moisture-testing equipment was part of the regulatory process. All because some mega farms can't seem to stop washing their dang romaine in contaminated irrigation canal water. Well, not just because of that. But small operators pay a disproportionate price for flaws in the mass agriculture infrastructure, and for the worldwide homogenization of trade.

Both the food system and its regulatory framework are designed for large, globalized operators, in cost and paperwork. One might accede to these kinds of rules if people were actually getting healthier as a result of them. As Edgar Winters puts it, "All the paperwork, it's got nothing to do with how healthy food is. All the tests we gotta do just for the State of Oregon alone, man, it's out of commission."

Within the Hemp in Hemp partnership, we're just beginning our research into Arizona and New Mexico food-grade production laws (and our introductions to the people with whom we'll be dealing in various government agencies), since that's where we'll be processing our coming hemp harvests. Colin is further along on the food-grade front with his products. He said Vermont's rules are pretty workable.

"We apply to the Department of Health, we prove that we used a certified commercial kitchen, we keep good records, and that's pretty much all there is to it," he said.

Vermont's is almost a cottage production regulatory framework mapped on a professional production level enterprise. Which is good. And worth noting that you don't hear many headlines about food contamination emanating from Vermont. But even Colin added that, in anticipation of coming federal hemp standards from both USDA and FDA, "We're working with in-state food safety experts who know which tests to run for our products. Best practices stuff. Like will the products mold if they have water activity, or is the pH of the products enough to make them shelf-stable?"

Nationwide, and collectively if we're wise, our strategy is to expand the cottage food concept to a wider professional industry category. This category will reflect the key role small-batch, top-shelf branding already plays and (let us hope) will continue to play right from the launch of the modern hemp industry. How should our regs read? Herein resides the crossroads (I won't say collision course) of what I think of as the Friendly Fungus with what we might call the Hairnet Era.

I first caught wind of the debate two decades before I ever imagined I might become part of it. It was when I was invited to taste an intentionally moldy rum in 1993. I was on a reporting assignment in Suriname. On a humid tropical evening in capital Paramaribo, I was offered a taste of a popular local offering. Iced. To procure it, folks went to the distillery and personally refilled their bottles from aged oak casks. Really aged. Like two centuries aged.

Normally, I'm not a fan of any hard alcohol, but this rum had an unbeatable Supermarket Pastoral. Plus, when in Rome, right? I allowed myself a measure from my host's cobalt blue bottle, and to my surprise the mixture tasted like liquid butterscotch. It was unforgettably delicious and smooth. Somewhere deep in my boxes of analog notebooks I have this rum's name written down. I think I had visions of becoming the exporter. At the time, on that colonial Dutch–era terrace, I inquired aloud of my host, "Why is this not the most popular drink in the world?"

The answer was a human-microbe relations issue. "It tastes so smooth because the molds that start the fermentation of the sugarcane have been building up on the casks for two hundred years." I was told. "International food rules require bleaching of production facilities. Then introducing new molds. Doing that would kill off what's special about this rum."

I like living foods. I eat as many living foods as I can. Furthermore, for most of human history, the kind of production mode I enjoyed in Suriname was the norm. In the words of author Stephen Buhner, "Yeasts have had a relationship with humans since our emergence on the planet."[2] The molds in the Surinamese rum's casks provided its taste profile. This is what I'm shopping for whenever possible, in all my food: the most long-standing production modes. If they didn't work, they wouldn't have endured.

No one wanted to die from tainted food even "back then." Maybe more did die. Maybe not. Evidently botulism in Alaska increased significantly when Western food methods replaced traditional Yup'ik fish fermentation techniques.[3] The point is humans had built-in techniques for food safety, or I wouldn't be writing these words and you wouldn't be reading them. Buhner devotes a fair amount of space to emphasizing our "special relationship" with the sugar-eating fungus *Saccharomyces*, an "integral part of the human diet for Millennia."[4] Kentucky brewer and author Jereme Zimmerman writes that medieval Scandinavians went so far as to save wild yeasts on "the juniper branches they used to filter their brews."[5] Even today, high-end vintners make use of the fungus called *Botrytis cinerea*, also known as *noble rot*, in the development of sweeter wines.[6] Ask a champion winemaker to wipe out the local wild *Botrytis* and she might weep.

Without these living critters, as Buhner describes them, we wouldn't have beer and we wouldn't have bread. Indeed he and Zimmerman argue, convincingly, that yeast is the first human-domesticated species. *Living* is the operative word here. As in our soil discussion, we're talking about complex, beneficial symbiosis with millions of other species in our products. Or at least harmless interactions.

When it comes to hemp, we're not necessarily referring to any one microbe. We're talking more about a philosophy of raw or at least minimally processed food. In the end, there are microbes and there are microbes. As Michael Pollan puts it, "Some of my best friends are germs."[7] When we speak of beneficial microbes, we mean beneficial not just for the soil but also for the garden that is our gut. This is why I never look for "anti-microbial" products. I look for microbial balance. In many cases, you want the good ones. Waging a war on all microbes is the thinking that causes superbugs. By crafting and eating living food, we're trying to nurture the ones that play nicely with us.

Obviously, there are dangerous microbes that we don't want in our food. Moving to the hairnet side of the discussion, in particular where it intersects with hemp/cannabis, I hark back to a 2018 conversation I had with a very smart Dutch colleague of mine, Sander Sandee. After establishing his reputation as a top-tier cannabis cultivator in the Netherlands, he developed an indoor food production protocol as part of his graduate work. We spoke backstage at Amsterdam's 10th annual Cannabis Liberation Day festival, where I was speaking and he was volunteering. He was leaving his native Holland the very next day for a career-track gig running an indoor cannabis grow outfit in Colorado.

In fact, we spent some of our fairly intense conversation that day hauling big pots containing his beautiful (and, I think worth noting, outdoor-grown) personal cannabis plants. They served as set dressing for the event's VIP area. Also they were in need of adoption, given Sander's pending emigration. He had brought the plants via boat, on Amsterdam's canals. They all found homes.

I want to stress that I don't think that following the philosophy Sander's work represents is wrong. He's a guy I respect. And if I ran an indoor grow facility aiming for certain types of compliance, he's the guy I'd want to hire. I just don't think it should be the only mode.

Sander is a fellow who can make sure every manner of abbreviation appears after your product's name. His shop talk includes QMS (quality management systems), MMRs (master manufacturing records), ISO (International Organization for Standardization) Quality Standards, and

GMP (good manufacturing practices). He singled out HACCP (hazard analysis and critical control points) as a food safety hazard reduction guideline that he thinks is important for every "grow." That's a noun that indoor cultivators substitute for *farm*.

Others in the industry recommend a Preventive Control Qualified Individual (PCQI) certification. This is evidently required by the Food Safety Modernization Act (FSMA), a 2011 federal law in the United States, drafted by large grocery trade groups, that expands FDA powers in controlling how food is grown and prepared. Parts of FSMA put forth exactly the kind of requirements from which we craft producers should be at least partly exempted. Or else we might have to float our own counter bill, let's call it the Food Safety Antiquating Act.

Those FSMA sections that craft hemp producers might want to further examine include forcing federally approved labs upon farmers, allowing the feds access to company plans, and allowing unannounced FDA inspections during which facilities and product are swabbed in order to make sure they are microbe-free. Our policy work will argue that a production facility should be "bad" microbe-free, of course. But kill all microbes in order to be allowed to provide our products to customers? Not so much.

Uniformity and clean grow environments are Sander's big talking points. This is the side of food-grade production that sees edible products as the end result of a "process" that needs "control."

"What does a clean grow environment mean to you?" I asked him.

"Effectively zero microbes," he said. "Automated clean rooms with no people inside."

So for one part of the hemp/cannabis industry, *clean* means "sterile." No 200-year-old molds in any 100,000-clone facility that Sander manages. That stuff will be taken out immediately. Also gone will be farmers touching every plant almost every day.

Guys like Sander are in demand. Anytime there's a romaine scare, they get more in demand. Sander provides a company with a paper trail that says folks won't get salmonella from their clones or isolate. This is the side of the industry that wants to push the plant as a pharmaceutical-grade product. Which is one way to go. It has a place within hemp's big tent. Until the orange-bottle generation moves on, some people are going to want pills dropped into one by a guy in a lab coat. Maybe the orange bottle can at least be made of hemp.

But for the independent farmer-entrepreneur to thrive, it's equally important that a whole-plant option, with all its nutritive properties still active in it, is also widely available for everyone, anywhere who wants it. Even when said plant is grown under that messy sun and in soil full of—shudder—fungus and other microbes. Ideally, we'll have both nutritive supplement and pharmaceutical options in the edible hemp marketplace. You can pick. I've heard lawyers refer to these two evolving branches of the hemp industry as two equal "streams."

In other words, you can today choose to eat a lot of fresh carrots, or take beta carotene pills. A level playing field in the hemp marketplace means if you want a cannabinoid product, you can buy a flower tincture provided by regional farmers or you can buy a pill grown in a grow facility . . . somewhere.

Erica Campbell, who is the farmer liaison for Senator Sanders and has visited our Vermont hemp fields three times, told me that her office is concerned about this level playing field. "Politicians are looking for constituents and stakeholders to make some noise and say, 'We want to make sure hemp remains a nutritive supplement, not just a pharmaceutical,'" she said.

Once again leaping into the phone booth to transform into Mr. Reasonable Middle Ground, all I'm saying here is that we must find a balance that protects hemp product safety without turning every sellable product into the cannabis equivalent of irradiated milk. That's because when you nuke any living product, an entire market sector and I believe, you potentially damage some of what is inherently beneficial in the product itself.

I'm the kind of guy who wants to see all the plant's cannabinoids, terpenes, and bioflavonoids, in their intended ratios, end up in the bottle. Living, organic hemp must have an equal if not predominant seat at the table. It's not just our brand. It's the healthiest brand.

The enduring predominance of our craft sector past hemp's launch phase will likely hinge on two things: (1) our effective organization as farmers to craft and enact the kind of regulations that will allow us to thrive, and (2) enough people making the decision to support local and regional enterprises.

When both of these happen, we've got ourselves an enduring market share. We'll be equipped to play in the big leagues alongside Big Food. In fact Craft Hemp will gain on Big Hemp the way Craft Beer is slaughtering Miller.

We will have our own lawyers and safety experts to help us negotiate key points based on craft beer and cottage models, but expanded to our 15-ton-or-less industry sector. And we're going to need them. In the coming debate (I won't say battle), we're also going to benefit from grown-ups in the room, like Roger Gussiaas, who see the significant and fundamental role of the craft market as hemp explodes back into the world marketplace. To have existing food professionals in our corner is essential. Roger's hemp-pressing operation might be the most advanced in the United States. It would be easy for him to focus only on the largest suppliers, bringing him the varieties that are the quickest to cultivate. Ya know, Mazola-style. Yet he has a place in his heart—and in his presses—for the artisan producer.

"Everybody wants consistency at scale," Roger said, echoing Chad Rosen. "But does that leave room for new flavor profiles and niche markets? I hope so. I'm willing to custom-process for those markets."

We can let the CBD mills that aspire to supply CVS fight over the kind of protocols that will result in ultrapasteurized hemp. If you're a raw food–leaning family like mine is, you should have commercial access to it even if you can't personally own goats or grow hemp. Our craft hemp regulations must allow a living-food middle ground. Best practices? Yes. Sterile hemp? Not for the craft stream of the industry.

It's an admirable mission, working to ensure a place for living products in a globalized food system. Will it succeed? In truth, we soil-based entrepreneurs are all too new at this to yet know if our values can map to the digital-age economy on a decentralized but cumulatively mass scale. The goal is mainstream buying habits demanding our products. Just as today every customer understands that a drink might come in regular and diet, so our endgame is a shopping climate wherein every purchase decision is either "regenerative" or "other."

Also exciting is that this is about more than one crop. If we succeed, the regenerative mode can dominate agriculture and the wider economy in the coming decades. When hundreds of millions of people start to shop this way, a thriving network of regional economies can succeed. As we rally customers, we're going to jujitsu our way into a seat at the table that carves out the industry's parameters.

Helpful Memos from Public Servants

We do tend, in our public policy, to be reactive. Partly, that's because any regulation emanating from one building along the Potomac that is intended to encompass 300 million stomachs is likely to incorporate a lowest common denominator. And now we're talking about a sort of planetary NAFTA of food regs in this Global Food Safety Initiative.

As the security expert Bruce Schneier emphasizes, many safety rules across society are intended much more to help masses of people feel safe than to actually help them stay safe.[8] Just as removing our shoes and dumping our water at airports doesn't necessarily make our flight more secure, so food regulatory regimes that bleach everything into sterility aren't necessary better for long-term health. We're arguably safer from salmonella, but we risk throwing out the baby with the microbes.

To give one example of misguided regulatory emphasis, during the first six months that I was researching the then illegal $6 billion ganja industry in the heart of the Emerald Triangle in 2011, there was exactly one federal raid in Mendocino County, California. Any guesses what it was for? Hint: It wasn't cannabis. It was raw cheese. Just for some perspective, between 2007 and 2012, there were no deaths reported from raw milk, while just under five people per day die from tainted meat, according to statistics from the Centers for Disease Control and Prevention.[9] No one seems to be screaming about that. It's just how the Big Food regulatory game rolls.

Anyone in a food business can give you concrete examples of safety protocols designed more for large operators than for independents. My local commercial raw dairy owner, for instance, Ashley White of Proverbs Farms, said she got around cost-prohibitive rules that, technically, required her to build a restroom solely for the visiting inspector of her five cows, by explaining that there was a perfectly serviceable bathroom across the street at the local airstrip.

Lest one think that initiating a craft market sector rally for friendly hemp regulations is a bit wonky and premature, I'll share that I'm still reeling a bit from a message I received on the day that the hemp-legalizing Farm Bill was signed. It was December 2018. I was just starting this book. While I was trying to take a minute to savor the most profound policy correction in three-quarters of a century, Cary Giguere had to drop me this line.

The subject read, "Of course there's this." I knew right away that inside would not be another of the "We win!" messages I had been getting and sending all day. I clicked the note open cautiously, wincing slightly. Before he eases my blood pressure, Cary often increases it. He presents problems, then solutions. What the note contained was a letter from then FDA commissioner Scott Gottlieb. It read in part:

> Today, the Agriculture Improvement Act of 2018 was signed into law. . . . Congress explicitly preserved the Agency's current authority to regulate products containing cannabis or cannabis-derived compounds under the Federal Food, Drug, and Cosmetic Act (FD&C Act) and section 351 of the Public Health Service Act. . . . In short, we treat products containing cannabis or cannabis-derived compounds as we do any other FDA-regulated products —meaning they're subject to the same authorities and requirements as FDA-regulated products containing any other substance.

I remember thinking, *That was quick.* I found it interesting that hours after a hemp provision representing 3 percent of the word count and even less of the budget outlay in a nearly trillion dollar agriculture bill was passed, the office of the FDA commissioner took time to instigate a pissing contest with farmers and entrepreneurs.

In many ways, the letter was the bureaucratic version of a playground bully's turf war—in this case with the assertion that one of the world's longest-utilized plants was now under FDA purview. Obviously the letter had just been waiting for someone to hit the send button. This was not the "Thank you, hemp pioneers, for helping the economy and climate!" one might have hoped for from our public servants. It was not, for instance, the attitude about hemp one sees in the current world hemp-acreage leader, China, which is, "Do it." Then president of China Hu Jintao paid an official visit to the nation's first modern-era hemp processor back in 2006.[10]

Ah well, it's just food, soil, and survival. But the key takeaway is, the time for craft farmer-entrepreneurs to get organized is now. The devil's in the regulatory details. This is true on both the state and federal levels. If you're an independent farmer, please participate in crafting regs. Personally. Not just by sending a check to an organization.

Here's how urgent this effort is: The FDA commissioner, running a department responsible for dealing with everything from opioids to food coloring to nut allergies for a third of a billion people, was so worried about CBD label claims that he railed on that theme for a couple of paragraphs in his note.[11] From Gottlieb's concern for our collective safety, one wonders how on earth humans managed to survive while eating this plant for 12,000 years. The letter was a wake-up call for everyone trying to help launch the industry.

Edgar Winters also saw the shot across the bow embedded in Gottlieb's note. His was one of my next calls that December day, and, of course, he expressed it most poetically:

"Hope you don't think this is the end of the battle," he said. "Anything that you can congest will be set by the FDA."

At least we know Cary is on the case, crafting the best possible federally compliant standards for the small-acreage farmers who dominate his state's farming economy.

"We have experience with this in Vermont from regulating the maple syrup industry," he told me as my circulatory system once again calmed into normal pressure ranges. He's just doing his job: Craft producers are a Vermont regulator's constituency. And here's some further encouraging news: Since ours is an industry already generating close to a billion dollars annually in North America, FDA policy honchos immediately felt public pressure following that initial letter.

"Since we put out that statement, this is one of the top issues I'm being asked during my visits on Capitol Hill," the FDA's Gottlieb said two months later, shortly before his resignation.[12]

Many months later still, as this book goes to print, everyone's on pins and needles waiting to see just how moderate or horrible initial FDA regs are. Will it be difficult or easy for whole-plant producers to provide product? You'll know the first federal regulatory scheme when you read these words. Regardless, the FDA salvo should keep us on our toes, looking those three turns ahead.

Overall, that memo should bolster our confidence: The tables have turned. So recently freed from the Controlled Substances Act, our plant is now one of the fastest-growing industries on the planet. Nothing has so impacted the world economy since tech came on the scene. Wear that knowledge with fortitude and pride as you lobby and educate. The Farm Bill was just the starting gun. We're defining the race. Be professional as you do your part to carve out that sweet spot between the hairnet era and our friend the fungus.

It's also helpful to remember that hemp is just another crop. Officially. Many people worked hard for decades to help it become recognized as such. You can shout this constantly: We're just farmers growing a plant. That's important for regulators to hear.

Taking in this weird initial landscape, the clever craft hemp entrepreneurs are planning for initial federal regulatory frameworks even while endeavoring to reform them as necessary to fit the regenerative mode. For her part, Janel at Palmetto Harmony is vocally in the trenches in the effort to ensure that whatever is defined as "best practices" will match what is required for a thriving independent model like her company's. Even so, like the Nohls in Vermont, she's investing in compliance in advance of coming protocols.

"We're working with a good manufacturing practices consulting company in anticipation of FDA standards for dietary supplements," she said, walking me into one of her "clean rooms." "We have two rooms that are up to FDA standards—one is our extraction facility and the other is our bottling, packaging, and R and D lab facility."

For producers, patience and persistence are again keys in this early regulatory phase, just as they have been at every stage of the hemp season. I was reminded of this the other day upon learning that a third-year enterprise—an ambitious, multistate processing effort—was giving up the ghost due to lack of capital.

Even when forced to drift into the policy realm, I find it's the plant kingdom reminding me to have faith in pace and dogged determination. The early season is my favorite part of the whole hemp cycle, laying down drip lines while sniffing the first apricot blossoms that have emerged overnight. I enjoy farmwork so much more than the regulatory issues we've been discussing. The moment you're in the field, you hit pause on the temporal, because you're connecting to the eternal. In fact, cartwheeling around my hemp field with bees is the way I want to spend my time *after* material success. But, given that I still have an entire bottled harvest to market, I have to think about these other things. As Margaret's Law reminds us, there is no part of the process the modern farmer-entrepreneur can ignore. Just as we have a long season of farming ahead come spring, so have we an intense year of policy work to tackle.

Which is to say, I've got my eye on the prize. I'm just allowing myself to play the dogged tortoise in my entrepreneurial work. Which feels good, from a secreting-of-happiness-chemicals perspective. I'm cautiously optimistic about the payoff. Both personally and for the wider regenerative craft industry. If we stick at it and minimize hypocrisy in our processes, I think many of us who identify ours as regenerative enterprises will thrive. We stay genuine most effectively by consciously connecting the dots in our work and our lives, even as we create our brands according to the Gospel of Dolly.

The "connecting the dots" concept is very simple: Live your brand. Don't phone it in with half-assed greenwashing or, if you're an investor, think of it as one of your purely absentee financial ventures of the 20th-century kind. Happily, regenerative farming and production and regional distribution appear to be the best routes to bottom-line success for the long term. Our business practices are our killer app. It's what we do that's phenomenal.

EPILOGUE

The Regenerative
Entrepreneur

To be happy at home is the ultimate result of all ambition.

—SAMUEL JOHNSON

D uring my conversation with the Vermont trooper at processing time, he'd asked me what I do for a living. A lot of
people wonder. Some guess Amish carpenter. Some guess
grunge drummer.

"Soil farmer," I said.

"Soy farmer?"

"Soil. Soil building—I work for healthy yields in crops."

He looked thoughtfully at my New Mexico license, then at the 50
pounds of cannabis flower in the backseat, then at the icy crust of snow
he'd just stomped through to reach my rig.

He said, "You must really care about soil."

It occurred to me a few hours later as I crunched to a former Vermont
girls' camp kitchen for the 13th time, just so I could process hemp by a
time-consuming method, that the trooper had nailed it. That's the starting point for connecting the dots in any regenerative enterprise. No one
has a problem with a promising bottom line—hemp, globally, is going

to be worth $13 billion by 2026.[1] But it's bottom line with a mission. At core, I'm trying to sequester carbon.

We who care all care for our own reasons. We all have a dog in this fight. I have a goat. I don't think it's an exaggeration to say that my family's lives have now been directly threatened by the climate situation. Over the course of the five weeks that the 2013 wildfire crept toward the four people and various furry critters for whose safety I am responsible, I slowly woke to the gravity of the mission.

Praying for rain in the most primeval way from his porch each morning as he watches certain destruction approach will start to awaken a fellow. Especially when he can't do a thing but pack. But it's when a terrified bear sends his family hustling for safety before mauling his goats in front of his eyes that it becomes personal.

I have to say I was pleased with the responses of every one of my family members to that "this is not a drill" morning. It being June in New Mexico, we had slept outside, in the big family tent we set up near the hammocks, between the ranch house and the goat corral. In the pre-monsoon heat, its mesh roof is a blessing, even before you add the New Mexico night sky. I fell asleep in a tranquil idyll of meteors and Pooh stories, and woke at dawn to galloping goat hooves and uptight chickens. This was not the normal start-of-day song.

The only human awake, I unzipped my window flap and migrated over 10 seconds from "as quiet as my mind ever gets" to "red alert." A large brown mass was scaling our eight-foot goat corral fence like a ladder. Before another 20 seconds had passed, we had our game plan sketched out and executed: Sweetheart brings human kids into house, then looks for gun case key (we never found it). Self goes and makes loud primate noises such as one would make to scare off a bear.

I bounded down to the horrible scene of my friends being killed, worked into a froth. I shook the corral fence, grabbed and threw a shovel over it, and generally danced around. You know how it is when you're five feet from a black bear, screaming at him to let him know you mean business. It didn't work for about 10 minutes, which was when my sweetheart called me up to the house and suggested I drive our vegetable oil–powered F-250 over to the corral and lay on the horn. After that came a lull in the attack. We'd exiled the bear to a meadow just north of the ranch for what turned out to be 12 hours. We didn't know how much time we had, so my sweetheart and I hauled gear to the corral and tended to wounded goats, battlefront-triage style.

Natalie Merchant, who had given us so much milk and was my primary morning meditation partner, was hollering something awful. My sweetheart and I locked eyes and realized at that moment that our family's survival depends on mitigating climate change. We've been working on it ever since.

Connecting the Dots in Our Work and Lives

Your family's survival depends on tackling the climate reality too. But you probably know that. That means, sure, cultivating hemp and sister crops widely, or passionately supporting those who do. That's a major piece of the puzzle, and one of the fun parts. The puzzle's instructions are, "Figure out how to work regenerative principles into the immediate, mission critical mind-set of all industrial enterprises on the planet, while educating the mainstream customer to seek the resulting products."

I don't think it's too much of a leap to plan for a broadly aware, farmer-centric buying public. Already there is widespread consumer awareness, for instance, that factory chicken eggs, produced by ill-treated animals, harvested by indentured farmers, and shipped thousands of miles, aren't actually cheaper than local organic eggs. Our task is to map that awareness onto hemp grown outdoors by regional family farmers.

Yes, a wartime mind-set is required, even though the work is peaceful. And everyone needs their personal climate Pearl Harbor. It looks like it won't be necessary to drop a terrified bear into everyone's backyard to get them into the unified spirit. The constant stream of fire, flood, hurricane, and tsunami chaos may well prove a catalyst for the sufficiently seamless migration to a regenerative economy. If the agricultural philosophy and techniques we're spearheading catch on, if the carbon gets sequestered, the plastic crap replaced, the batteries renewable, maybe we're in time.

There's no way I could let up, even if I wanted to. The 2013 Silver fire, as the Forest Service calls it, still resonates. I have to drive through its skeleton every time I leave the ranch for the wider world. For miles upon miles, entire hillsides are torched but impressively recovering. Starting about last year, the fifth anniversary of the blaze, I began seeing a wildflower palette spread across the char. This spring, as we hike at high altitudes, we can trace the next stages of a ponderosa pine forest cycle: mountain mahogany, three-leaf sumac, and, along the creeks, alder, all doing the same thing that you and I are: building soil.

A forest ecosystem cycle is, like all systems that have endured for tens of millions of years, a regenerative one. Both fire and a tree's response to it were programmed into the system long before humans entered the game, thanks to lightning. The forest has been to this fire carnival before, and it normally takes 250 to 1,000 years to recover from a fire like the Silver fire.[2] The difference is that lately (since coal burning), humanity's systemic choices exacerbate routine cyclical events into millennial ones.

In the Southwest high desert, our once reliable monsoon season would provide 40 percent of our watershed's annual precipitation. Daily afternoon thunderstorms are supposed to begin in early July and peak in August. I'd say that happens every third year lately. Old-timers tell me my own ranch's creek used to run nearly year-round. Now it runs maybe a month out of the year. It's an event when it does. As in, "Creek's running! We'll pick up this math lesson later!"

September is our spring—a wildflower dreamscape rises from the arroyos and buttes: It's magnificent. You haven't lived until you've smelled a desert willow in bloom or nibbled on a lemonade berry, or watched a sphinx moth get drunk on datura nectar at midnight. Now, with our seasonal arc disrupted and erratic, extended dry conditions allow bark beetles to weaken millions of the majestic, fragrant ponderosas that surround the Funky Butte Ranch. The trees—some three centuries old—die where they stand, leaving mountainsides of brown-needled fuses, waiting.

I recognize we might be getting numb to these kinds of stats, but as I type, six years after the fire, 2019 has been the second-hottest April in 139 years of record keeping, according to NASA.[3] Carbon was measured at—ho hum—415 parts per million that month.[4] First time in three million years. We're number one!

Beyond the morning the bear came to breakfast, this climate change is not academic for me. It has changed life on the ranch forever. Some mornings when I stumble half conscious to the corral for milking, the hair on the back of my neck stands while I snap into a sort of ancient watchfulness mode for a few moments, like a dog sniffing the air.

We all want our families to be safe, healthy, and comfortable. That's why we work. Doesn't matter if you're a farmer or a real estate attorney. The issue is whether you're looking ahead one fiscal quarter or seven generations. Since giving a dang is the secret ingredient in this industry, I'd like to petition Ms. Parton to add this postscript to her mantra: In addition to First, Better, or Different, you've also got to be the Most Unflinchingly Dedicated to Achieving Your Calling, Which Calling Is Tied into the Greater Regenerative Mission. Or something more concise. Maybe, "First, Better, or Different, plus Kind."

I'm setting it down here for the record, and I guess we'll see: A regenerative model is going to win in the coming economy. Folks who insist on what we wrongly today call "conventional" farming or wider industrial techniques, I predict, are going to be left in their own toxic dust. They won't be able to compete, in productivity or quality.

Humans experimented with unrestrained harvesting of our divinely given resources. We saw that it didn't work. Easy fix. At every level of society and economy, as individuals, parents, entrepreneurs, investors, dinner eaters, we're shifting into post–Pearl Harbor single-mindedness. Ninth inning. Two outs.

Connecting the dots starts at home, because everything starts at home. Here on the ranch, we milk goats to reduce the carbon miles in our milk, yogurt, and cheese. But twice in the past dozen years we've scaled down the herd to allow us to travel together more often. That means buying dairy again, albeit from neighbors.

What can we do besides try to improve, to keep edging forward with baby steps? The journey is a life- and career-long one, and, on the business side, too, no mortal is at the Promised Land unless her delivery vans run on solar-charged hemp batteries and every farmer in the operation is living the worry-free good life. Myself, I've long been engaged in a personal hypocrisy reduction project. I stand ready to be educated in the inevitable areas where I or an enterprise of mine can take it up a notch. Man, solar-powered electric commercial air travel sure will help my carbon footprint.

All this consumer-side stuff—from righteous product labeling to not ordering company T-shirts made by Bangladeshi children—is going to seem so laughably obvious in 20 years. It's a modern rite of passage: One day we all wake up and realize, wow, all my stuff was made somewhere—I wonder by whom. I wonder what's in it.

Carrying that awakening to the production side, what we regenerative hemp entrepreneurs offer is such a change in our manner of interacting with the earth that simply demonstrating our principles in our products embodies a successful strategy at this stage. This is why I would love to see as many of us as possible succeed—to still be in it and living well and sequestering carbon in seven decades and generations. That's when 30 percent of Americans will be making a dentist-neighborhood living from independent farming.

For the moment, we're still an agricultural and industrial blip. We have seven billion people to feed, house, and clothe. So every hemp sale is a sale for my enterprise and yours, probably for five more years at least. Maybe 10. Collectively, we're a plane flying a banner over a crowded beach reading, HEMP/CANNABIS IS OPEN FOR BUSINESS, EVERYONE! ASK AROUND! KNOW YOUR FARMER!

If you worry about facing judgment for leaping in unprepared, then start small. I love it when someone tells me, "We've got a baby on the way, we just got out of debt, so we're going to start with three acres. But we have twenty available for next year." Just start.

As Jeremy Fisher, a 25-year-old first-time Arkansan hemp farmer who helped write his state's regulations, put it when he noticed his three-acre field was, like much of Arkansas and Oklahoma in the spring of 2019—ho hum—under five feet of flooded Arkansas River, "You don't see Pop Warner football players leaping directly to the NFL." So much for plan A. That first crop might not look stellar, but I'd be willing to wager that Jeremy will still be planting hemp in five years.

Examining the near future marketplace with a wider-angle lens, a real sign of victory will be when those "leading economic indicators" you hear evaluated on the top-of-the-hour NPR news updates no longer catalog "new home starts" or "cost of the latest bailout," but rather "inches of soil rebuilt" and "number of regenerative agriculture sector jobs created." Then, maybe, millennial events will occur once a millennium again.

Lollygagging as a Family Practice

Representatives from the valley's honeybees and at least four species of native bees circled closely around us as my family and I dropped the first seeds into ranch soil at the end of May 2019. We gave a formal prayer of thanks. The dogs panted patiently under a walnut. The time-lapse camera was rolling.

Planting our first home hemp crop was a very big moment for us. This is where our sons were born. But it was the sheer number and variety of bees that kept grabbing our attention. They ranged from housefly-sized to small delivery drone. We talk about saving the bees, but I didn't realize how fast it would happen. The blindingly bright leaf-cutters kept tunneling their homes all around the ranch house. As folks do with turtle nests in the tropics, we set up warning signs and danced around them.

And as with the turtles on a resort beach, we kind of wondered, *Why right here?* These solitary native bees are surrounded by two million acres of wilderness, so there's something they like about raising a family close to us, even with wrestling dogs and goat kids and human kids. I think they might feel they are taking care of us. Maybe they have a Save the Humans campaign going. They sure work hard in our garden.

We love watching them burrow back home after a day drinking catmint, horehound, lavender, and bee balm. And, any day now, hemp. We keep trying to film them digging in slo-mo.

The prayer done, we started working around the field, hand-planting. Almost immediately I found myself in conversation with the local rabbits, of which the 2019 batch was stereotypically prolific. My position was basically a standing invitation to treat this crop more as an appetizer than a main course.

"We're over-planting," I said evenly, my mouth pressed close to one of the two softball-sized warren holes that had already opened inside our half-planted field. "Just so you, the birds, and the grasshoppers can have a taste. We request that you treat this year's hemp as a sometimes food. So there's enough for everybody."

This crop, if all went well, would provide a good portion of our human and goat protein for a year. Zero carbon miles. Only good additives. And very few of those. I was trusting the rabbits understood the concept of pace. It was a small crop, under half an acre. I'd rather not have to go through the song and dance of circling each plant with chicken wire.

It also became clear in the days leading up to planting that our hemp would be sharing soil space with the native vegetation. We were, in fact, unintentionally companion planting with spectacle pod, horehound, and even—shocking as it would be to Dan Townsend—mullein. So far it was working out fine. Seemed like there would be room for everyone, as I shook seed into my loved ones' palms.

The effort at interspecies cooperation set me thinking for a moment about the squabbling we're starting to see in the cannabis farming community, among fans of one or another of the supposedly different sides of the plant. Obviously, as we've been discussing, throughout nearly all of humanity's relationship with this plant, there have been no different sides. It's always been one plant, no matter how you chose to grow it, and no matter what your intended final products. Formal THC irrelevance will make that return to equilibrium official. Communication is what's required among different styles of cultivation, old and new, and among various desired cannabinoid harvests. I mean, if rabbits can grasp this . . .

I realized with a laugh that in my discussion with the Funky Butte Ranch leporids I was employing the same advice I give to farmers growing the hemp/cannabis plant for any application: Follow the example of farmers throughout the ages. Call your neighbors. Communicate. Bring your work gloves to one another's farms and learn about one another's techniques, dreams, and kelp sourcing.

I breathed deeply, exhaled, and let non-field issues escape through my muddy fingertips. It was easy to do. I got back to business. We were approaching June. Spring training was over and the real season had begun. Records were being kept, which ate into cartwheel time. I try to allow for about an hour of bureaucratic BS per day.[5]

18 Days Later . . .

The first thunder of the season, rumbling in a long diminuendo, hints at a functional monsoon. Two ravens court in a graceful daredevil routine not far overhead as I and my kiddos water at dusk. I'm taking a brief break to make these notes:

The family crop itself is small enough that I know the quirks of each soil area; no FLOATER delays in this field. In my twice-daily meander through the field, tucking in keikis and tweaking drip lines, I recognize every plant. We each already have our favorites. As the desert cools this evening, I report joyfully on the progress of Magu, named for an ancient wellness goddess associated with hemp across Asia. My oldest just shouted, "First serrated leaf on the Purple Wonder Twins by the currant bush!"

These keikis are strong; able to stand up to searing midday desert heat (they love the sun, and follow it devotedly), and to ignore some rabbit nibbles. Once again the plant is showing me that there is nothing like the strength of hemp sprouted from seed in outdoor soil.

Based on their robustly dark olive-green stems and rapid growth, they seem, like the bees, to feel very comfortable in their homes. Which is important. The top three keys to contentment in life, the Realtors correctly advise us, are location, location, and location. I'm no exception. I'm in the soil with the people I love.

There's something about farming that mitigates the profound neural changes that have been brought about by our interaction with zeros and ones. Or maybe even since writing. The desire to ranch and farm stayed with me through four generations of city life, suburbia, and a transatlantic steerage voyage. It is who I am, at a fundamental genetic level. It's where I am healthiest.

My family's life is centered on getting our fingers in the soil, every day. Tomatoes, eggplant, elderberry, hemp, currant, mulberry, locust, broccoli, onion, corn, and arugula. My own fingers are browning the phone's keyboard as I type. Seems like our only hope, to live this way. I can't say I'm sure of the odds, but it seems worth a try at least. Beats the pants off the alternatives.

Watering and weeding the hemp field now constitute my only waking-life writing breaks—this book is due in 10 days. I extend my tasks here as long as possible. I can't get enough of the sudden cicada crescendos, the complex canyon wren trills and, of course, the miracle of germination.

I use two fingers to buttress a keiki emerging from the goat-poop-and-alfalfa compost we've been building. In that motion, I accidentally brush the fuzzy back of one of the wild bees—a carder I think, they all look so similar in the insect book. An earthworm hulas a brief good evening before disappearing back into the microbe condo. Nobody gives the impression of being disturbed.

My eye is pulled to a green shimmer radiating glossier than any graffiti tag. Its nail polish name would be "Neon Flux." Ah, it's a beetle shell, itself suggesting a spot where I might start extending the radius I'm giving to this Samurai plant of which my sweetheart is particularly fond. (We've named her the Hemp Yeti.) She's bifurcating at the origin point of the hemp spiral we've planted in this place that means everything to us.

As I gently move a clump of compost in the spot suggested by the beetle shell, I find it. The white mycelia streak. A lightning bolt. A helix. A fractal. Population: 10 or 20 million, all living upper-middle class mycelial lives in the highest desert landscape you've ever seen. Nurturing this fungus is something that, regardless of any entrepreneurial result, feels positive for the cosmos. For another year, the bees are here, my family eats superfood, and the soil is alive. This is a feeling all humans had until recently, upon watching a crop come up: We will survive.

I scan for my sons, thinking I'll call them over to see that our spring fungus gathering has paid off. They are off examining a curved-bill thrasher nest in a cholla not even close to the hemp plants they are supposed to be tending. Strangely, I find that my greatest appreciation in this moment is for the endangered practice of lollygagging. It feels like a key part of the joyfulness woven into these past weeks in the field.

Most of us rightly subscribe to the pop science belief that emptying your mind, at least for a few minutes, resets the creative side for when you head back to "work." It's a neural yin-yang thing. We all do it, because it works. Iginia, she'll be delighted I'm disclosing, enjoys watching champion lip-synchers. I've seen Cary unwind by watching the Tiger Woods of snooker, Ronnie O'Sullivan, run the table. YouTube has been helpful to the modern version of this concept. But the value of zoning out was well known prior to mobile networked devices.

I love watching my sons trip out on leaves in a breeze, or whatever it is
that captures them for 6 or 7 minutes at a time. When we're quiet is
when we most often see a member of the fox family with whom it
looks like we're going to be sharing our hemp and tomato crops this
season. Plenty, plenty, plenty to go around. In the final analysis, that's
what regenerative living is: the rejection of "us or them." And of worry.

How did the stress and the overscheduling happen? Many, of course,
blame the mobile networked devices themselves. Regardless of the
cause, neural recharging is real. It's why I so appreciate its encoding in
a day of rest. Otherwise there's no end. It's always hemp o'clock somewhere. I've set some crazy alarm times for conference calls with Kenya
and Western Australia. That's when you see moonrise and sunrise.

Today is not one of those days. But more than a few recent ones have
been, and, if trends continue, many forthcoming ones will be too.
Armed with that awareness, I find myself wanting to end this story of
a season in regenerative hemp with the invocation that while you're
jamming to help humanity (including your kin) survive, you're also
enjoying the ride.

I hope you check yourself every couple of days to make sure you're prior-
itizing your family over all else, and even taking the odd afternoon off
to get on the lake or whatever. Because, as John Lennon says, "pretty
soon you're gonna be dead." To my human kids who are two of this
book's proofreaders, I'd like to officially say that, no, this praise of lollygagging does not mean that you can skip homeschool lessons today.

Unless, of course, it does. There's always a plan B. And where there is a
plan B, there is hope.

ACKNOWLEDGMENTS

The independent farming renaissance, like the early phases of most important social movements, is an awesomeness magnet. It is attracting some of the finest, cleverest, most dedicated folks I've ever met. And, of course, per the basic rules of physics, some doofuses. Just a few. As is my wont, I'll focus on the positive. And of course we all learn from everyone who crosses our path.

I couldn't plant, harvest, process, and market my own hemp, let alone write this book, without the following folks: Edgar Winters, Margaret Flewellen, and family in Oregon (including the genius Randy and next-generation superstars Chris, Celeste, and Dougie) are, as we've seen in these pages, true mentors and friends. Cary and Kristen Giguere, Colin and Erin Nohl, Fiona Giguere and Basil are all terrific colleagues and pards.

Since I've already been referring to them as kin, it won't surprise anyone that the folks at the Fat Pig Society organic hemp cooperative in Colorado (Iginia Boccalandro, Bill Althouse, Gavin Lim, Yamie Lucero, and the late great John Long) are guiding lights for me. They set such a high bar as colleagues and as friends.

Roger Gussiaas has been an invaluable friend, colleague, advisor, and confidant. Vincent and Irene Mina and family demonstrated for me how an independent farming family can make a living in a *very* small space. Dexter Rice of Nature's Love and Sub-Zero Extracts is one of the most generous folks I've met inside or outside of hemp. Andrew and Jacob Bish and the whole crew at Bish Enterprises are total Renaissance people and great friends. I'm grateful to Shane Davis of Boulder Hemp Farm and M222 Genetics for starting me thinking about Ogham culture and plants. Preston Whitfield and Louis Zerobnick are unwavering brothers.

ACKNOWLEDGMENTS

Thanks to Morris Beegle, Lizzy Knight, and Lori Buderus at WAFBA, who have supported my work, especially live performance and tree-free printing, for six years. Shelly, Erin, and team at the superlative Montana State Hemp Festival have been dedicated to allowing me to spread the farmer-first gospel in the Rockies. Same with Nancy, Amber, the two Laurens (Stansbury and Berlekamp), and all the great folks who have been organizers at Hemp History Week.

Special thanks to Hana Gabrielova, Blake Miller, Nashville Rizzi, Ken Manfredi, Robin Alberti, John and Melissa Williamson, Gram and Larry. I'm grateful to Agua Das for the regional lime he provided for my early hempcrete experiments (and for Hemp I Scream).

Big appreciation to all the hemp and independent farming folks with whom I continue to cross paths—from Kentucky to Belgium, I gain knowledge from every one of you, about everything from tractor maintenance to the joys of beeswax-vented greenhousing. Some folks in that list of treasured humanoids: Margaret, Aaron, Fran, Joe, Trevor, Liz, and the (scary) cows at Salt Creek Hemp are my peeps. Immense thanks to Jackie Richter for her belief in me from the start. She has one of the toughest jobs in hemp, and I think she is going to succeed. Thank you to Melody Heidel and Qing Li for giving me the subtropical hemp bug. Good thing: Equatorial belt humanoids have endocannabinoid systems too. I'm grateful to Andrew Stoll, Dean Norton, Charlotte Rosendahl (that early MOU saved me much sleep), Pauli Rotterdam, Bobby Pahia, James and Janell Simpliciano, Tomas Balogh, Dan Townsend, Dani Fontaine, Wild Bill Billings, Derrick Bergman, Mich Degens, Evi Royackers, Mary Bailey, Joy and Don Nelson, Annie and Willie Nelson, Micah Nelson, John Roulac, David Bronner, Anthony Johnson, Sarah Duff, Darrell Koerner, Dale Sky Jones, Andrew DeAngelo, Steve DeAngelo, Barry Gordon, Jake Gordon, Adam Eidinger, Andrew Stone, Sean Marks, Marios Rush, Ryan Slabaugh, Sarah Levesque, Chad Kuskie, Ben Trollinger, and everyone at Acres USA, Ryan Loflin, Ed Lehrburger, Carl Lehrburger, Eric Carlson, Rachel Cole, Marc Grignon, Stephen Jackson, Duane Ludwig, E. R. Beach, Tara Grace, Nami, Lelle Vie, John Delgado, Gloria Castillo, Jerry Fuentes, Bill Gomez, Lew Seebinger, George Rixey, Richard Dash, Ron Alcalay, Jason Amatucci, Josh Hendrix (thanks for early introductions to the Kentucky hemp family),

Ben Droz, Matt McClain, Annie Rouse, Arthur Rouse (thanks in particular for the Hale reference and for the green screen), Ellen Komp, Coral Reefer, Rick Trojan, Dale Gieringer, Michael LaBelle, Peter Nyari and the team at Hemptique (we use their hempen rope to corral goats and secure treehouses), Ewket Assefa, Melissa and Josh Rabe, Janel Ralph, David Newsom, Anson Tebbetts, Anthony Iarrapino, Victor Guadagno, KMO, Philip Ackerman, Caroline Kimball, Mike Gabbard, Tulsi Gabbard, Emily Emmons, Cynthia Thielen, Erica Campbell, Mark Frauenfelder, Dion Markgraaff, Chad Rosen (big thanks for letting our little craft pressing leap the queue on short notice), Shane Ball, Joe Hickey, Katie Moyer, Eric Steenstra, Colleen Keahey Lanier, Melissa Peterson, Marty Phipps, Mike Lewis, Shadi Ramey (who steered me to anthocyanins), Pamela Jennings Orth, Joy Beckerman, Bill Collins, Wendy Gibson, John Patterson, Kristen Kunau, Tommy Nahulu, Ben Wright (lovely design for the cover of the hemp-printed *First Legal Harvest* monograph), Liz Jackson, Kainoa Aluli, Lehia Apana, Brad Bayless, Rob Parsons, Chris Harris, and Kendal Clark. RIP Mark Linday, who was leading the hemp-plastic revolution and printed a lot of hemp goats for me.

Michael Metivier is a great editor, and it's a pleasure to be working with Chelsea Green again. Thanks as always to Markus Hoffmann, my agent.

Special thanks to the engineers and designers who made the 2013 MacBook Pro goat proof. This one might sound cheesy, but it's true: The cannabis/hemp plant's intelligence has guided me and is worth thanking.

Most vitally, as usual, the unwavering love and support from my family here at home on the Funky Butte Ranch allow me to do what I do. You are all that matters. Everything else is gravy. And we're finally growing hemp at home!

A few names have been changed in these pages to protect privacy.

NOTES

INTRODUCTION: REFUGEE BEAR

1. As you might already have discerned, all our Funky Butte Ranch goats are named after singers we like, but whose voices might tend toward the caprine.

CHAPTER ONE: BE FIRST, BETTER, OR DIFFERENT

1. Peter Yang, "Miners vs. Merchants: How Global Trade Made Men Wealthy During the California Gold Rush," *Flexport* (blog), May 3, 2016, https://www.flexport.com /blog/trade-merchants-rich-california-gold-rush.

2. Phil Buck, "Florida Cannabis Director: Hemp Will Be a Billion-Dollar Industry for State," *WTSP News*, March 7, 2019, https://www.wtsp.com/article/news/florida -cannabis-director-hemp-will-be-a-billion-dollar-industry-for-state/67-98b873 cf-4b6b-4061-86de-d4450a550ab0.

3. Ernest Small and David Marcus, "Hemp: A New Crop with New Uses for North America," in *Trends in New Crops and New Uses*, eds. Jules Janick and Anna Whipkey (Alexandria, VA: ASHS Press), 284–326.

4. M. Salzet and G. B. Stefano, "The Endocannabinoid System in Invertebrates," *Prostaglandins, Leukotrienes and Essential Fatty Acids* 66, nos. 2–3 (February 2002): 353–61, https://doi.org/10.1054/plef.2001.0347.

5. The only legal difference between "hemp" and "cannabis" is that the former contains 0.3 percent THC or less, however it's tested. That definition, as it should and as we'll discuss, will be changing.

6. Sabatino Maione et al., "Non-Psychoactive Cannabinoids Modulate the Descending Pathway of Antinociception in Anaesthetized Rats Through Several Mechanisms of Action," *British Journal of Pharmacology* 162, no. 3 (February 2011): 584–96, https://doi.org/10.1111/j.1476-5381.2010.01063.x.

7. Baily Rahn, "Why Does Cannabis Produce THC?" *Leafly*, December 16, 2016, https://www.leafly.com/news/science-tech/why-does-cannabis-produce-thc.

8. This theory has a counterpart in nutrition. Many experts believe that the importance for dietary absorption of essential fatty acids resides in the ratios of omegas 9, 6, and 3 in your food. And many nutritionists believe an ideal ratio of all three can be found in hempseed and hempseed oil and protein.

9. Brightfield Group, "U.S. CBD Market to Grow 700% Through 2019," press release, July 9, 2019, https://www.brightfieldgroup.com/press-releases/cbd-market -growth-2019.

10. Julia Rosen and Anna M. Phillips, "Land Use Policy Key to Reining in Global Warming, U.N. Report Warns," *Los Angeles Times*, August 8, 2019; "Land Is a Critical Resource," *Intergovernmental Panel on Climate Change Newsletter*, August 8, 2019.

11. Rodale's three pillars of regenerative organics are soil health, animal welfare, and social fairness. For more details on the concept's origins, see https://rodaleinstitute.org.

12. Judith D. Schwartz, "Soil as Carbon Storehouse: New Weapon in Climate Fight?" *Yale Environment 360*, March 4, 2014. https://e360.yale.edu/features/soil_as_carbon_storehouse_new_weapon_in_climate_fight.

13. In the psychoactive cannabis market, where THC is a primary goal, flower has always been the cultivation target.

14. Before my first cannabinoid test showed its small but distinct presence, I didn't know what CBC was. Once I learned about its possible muscle-relaxing effects—*poof*—I had a soothing massage and bath oil.

15. Plus, cultivating as they are in their backyard with a long-term vision, the regional enterprise's team will almost certainly prove more effective in our core soil-building mission.

16. Based on several studies aggregated by the American Independent Business Alliance, at https://www.amiba.net/resources/multiplier-effect/.

CHAPTER TWO: WE'RE ALL SOIL FARMERS NOW

1. Jean Nick, "The History of How Organic Farming Was Lost," *Eat Well Do Good Blog*, Nature's Path (website), December 30, 2016, https://www.naturespath.com/en-us/blog/the-history-of-how-organic-farming-was-lost/; "Roosevelt Urges States to Create Conservation Districts," *Rice Soil and Water Conservation District Newsletter*, Faribault, MN, May 8, 2017.

2. United States Department of Agriculture (USDA), *Soils and Men: Yearbook of Agriculture 1938* (Washington, DC: United States Government Printing Office, 1938), 864.

3. Daniel Cressey, "Widely Used Herbicide Linked to Cancer," *Nature*, March 24, 2015, https://www.nature.com/news/widely-used-herbicide-linked-to-cancer-1.17181.

4. Willie Crosby, "What Is Mycelium: Nature's World Wide Web," Fungi Ally (website), June 8, 2018, https://fungially.com/what-is-mycelium-natures-world-wide-web/.

5. Kelly Levin, "New Global CO2 Emissions Numbers Are In. They're Not Good," World Resources Institute blog, December 5, 2018, https://www.wri.org/blog/2018/12/new-global-co2-emissions-numbers-are-they-re-not-good.

6. "All About Grains: Wheat," Washington Grain Commission, http://wagrains.org/all-about-wheat/varieties-of-wheat/. USDA Foreign Agricultural Service, "Commodity Intelligence Report: Crop Production in Greece and Italy," August 18, 2017, https://ipad.fas.usda.gov/highlights/2017/08/greeceitaly/index.htm.

7. Kalo is a nutrient-dense "canoe crop," also known as taro, brought over by the original Hawaiians millennia ago. I think a great slogan for a revived traditional island food movement would be, "Nutritious enough to haul 2,500 miles in a canoe."

8. I earlier visited with Althouse in *Hemp Bound* when he chauffeured me in a hemp-powered limo (once owned by Imelda Marcos) while we delivered cannabis plants to veterans with PTSD and discussed biomass-based "gasification" energy futures.
9. Now I wear shades with frames of hemp and repurposed wood.

CHAPTER THREE: OWNING YOUR SEED

1. Ernest Small and Arthur Cronquist, "A Practical and Natural Taxonomy for Cannabis," *Taxon* 25, no. 4 (August 1976): 405–35, https://doi.org/10.2307/1220524.
2. Ernest Small, interview by Robert C. Clarke, December 1999, http://www.international hempassociation.org/jiha/jiha6208.html.
3. On November 8, 2018, the Senate's senior Republican hemp supporter, Mitch McConnell of Kentucky, promised hemp would be "lightly regulated" by the USDA. At this moment so soon after the Farm Bill's passage, we're in wait-and-see mode. Hearings are being held, and initial policy could be released just before or soon after the initial publication of this book.
4. Your state legislature passes the law, then your state agriculture department drafts the rules. I suggest you participate in the rules process. Pester your state hemp administrators. They love that.
5. Vote Hemp, *2018 Hemp Crop Report*, https://www.votehemp.com/wp-content /uploads/2019/06/Vote-Hemp-Crop-Report-2018-v2.pdf; Brittany Falkers, "Hemp Farming on the Rise in Oregon and CBD Business Continues to Grow," KGW8, June 19, 2019, 10:41 a.m. PDT, https://www.kgw.com/article/news /hemp-farming-on-the-rise-in-oregon-and-cbd-business-continues-to-grow /283-d290feeb-b85c-4ded-aa78-bce67b78be1a.
6. "Industrial Hemp Production in Tasmania," Agriculture, Tasmanian Government Department of Primary Industries, Parks, Water and Environment, https:// dpipwe.tas.gov.au/agriculture/plant-industries/industrial-hemp; Melissa Schiller, "International Hemp Industry Hopes to Change U.S.'s 0.3-Percent THC Limit," *Cannabis Business Times*, June 12, 2019, https://www.cannabisbusinesstimes.com /article/international-hemp-industry-hopes-to-change-us-thc-limit/.
7. Cold ethanol extraction is probably the most popular mode of flower processing today. A cold ethanol extractor is essentially a modern version of Uncle Jesse's still, except you end up taking the alcohol out of the final product (see "Cold Ethanol," chapter 10, page 170).
8. Testing protocol is one of the components of hemp policy that might be established by USDA policy when such policy is released. Let's hope it's farmer friendly (leaf testing for sinsemilla and no testing for dioecious crops). If so, great. If not, let's change it. To date, the main thrust of the chatter among state hemp-program coordinators is which of the two primary testing methods (gas chromatography or high-pressure liquid chromatography) will become the standard. There are arguments for or against both, but for our purposes this matters much less than establishing a friendly protocol like leaf testing as the standard for THC tests in your state.

9. Ernest Small, interview by Robert C. Clarke (see chap. 3, n. 2).

10. Kerstin Iffland, Michael Carus, and Franjo Grotenhermen, *Decarboxylation of Tetrahydrocannabinolic Acid (THCA) to Active THC* (Hürth, DE: European Industrial Hemp Association, 2016).

11. Renée Johnson, *Hemp as an Agricultural Commodity* (Washington, DC: Congressional Research Service, July 24, 2013), https://www.everycrsreport.com /files/20130724_RL32725_b2a91097ba168413aa8806f6bc5215fa20ee97de.pdf.

CHAPTER FOUR: WILD WEST GENETICS

1. You can identify gender fairly early in the plant's growth cycle, about five or six weeks after planting.

2. At least they start out identical, and I've seen them remain so under grow-room conditions. However, a week after I brought home my first four clones as a possible supplement to the ranch crop and introduced them to sunlight, they displayed four completely different growth patterns. Two were branching bushes, one was a narrow tree, and the fourth was much smaller and denser than the others. All of them kept me company in my office, sequestering carbon and adding that helpful extra bit of oxygen as I jammed toward completion of this book.

3. OMRI stands for Organic Materials Review Institute.

4. Bandana Chakravarti, Janani Ravi, and Ramesh K. Ganju, "Cannabinoids as Therapeutic Agents in Cancer: Current Status and Future Implications," *Oncotarget* 5, no. 15 (2014): 5852–72, https://doi.org/10.18632/oncotarget.2233; Francesca Borrelli et al., "Colon Carcinogenesis Is Inhibited by the TRPM8 Antagonist Cannabigerol, a *Cannabis*-Derived Non-Psychotropic Cannabinoid," *Carcinogenesis* 35, no. 12 (December 2014): 2787–97, https://doi.org/10.1093/carcin/bgu205.

CHAPTER FIVE: ADVENTURES IN PLANTING-GEAR MALFUNCTION

1. Calvin Hughes, "Hemp Is a Godsend for Bees, Says This Colorado Insect Expert," *Civilized*, July 20, 2018, https://www.civilized.life/articles/hemp-is-a-godsend -for-bees-says-this-colorado-insect-expert/.

CHAPTER SIX: ON WEEDING, WATERING, AND ORGANIC CERTIFICATION

1. At the time of this writing, David Bronner of Dr. Bronner's soaps, for instance, is spearheading a new, more comprehensive Sun+Earth certification for cannabis that figures in not just regenerative principles but fair trade ones.

CHAPTER SEVEN: THE MIDSEASON PANIC ATTACK: ROOKIE FOCUS

1. "Did George Washington Grow Hemp?" Facts, George Washington's Mount Vernon (website), https://www.mountvernon.org/george-washington/facts/george -washington-grew-hemp/.

2. The continent-wide wall of smoke during that trip made life a perpetual twilight, canceled Dan Townsend's son's football games (due to live aerial cinders), and left piles of ash that some old-timers called "deeper than Mount Saint Helens" in parts of Washington.

CHAPTER EIGHT: THE MIDSEASON PANIC ATTACK: VETERAN FOCUS

1. Midwesterners looked at me with puzzlement when I said the word *silo* at conferences until I learned that their regional term for this is *grain bin*. You know, the kind of thing the subjects of a Grant Wood painting stand next to, also known as the only topological feature in the state of North Dakota. Adding to the regional confusion, Cary and his fellow Vermonters call them *silage bunkers*. I am using the term *silo* to mean any container in which you can deposit your seed harvest so it can be immediately dried via an upward-facing fan mounted at the bottom of the receptacle.
2. We'll talk about how to tell when a seed crop is ready when we get to harvest.
3. What happens biochemically when you sniff hemp flowers is your body secretes anandamide, an endogenous cannabinoid discovered by Raphael Mechoulam and Bill Devane in 1992. It is also known as the Bliss Molecule.
4. We had an exciting unicorn display in Oregon that same season. I think the smoky air chemistry might be why about 15 percent of our fiber harvest showed a magnificent purple tinge. It was a shade that longtime hemp textile retailer Larry Serbin of Hemp Traders told me, when he saw me wearing a bracelet I had made from it at an industry convention, he had never seen before. "You got yourself a unicorn there," he said, possibly due to the variety's anthocyanins, or vacuolar pigments. "I'd like to see more of it." It was presmoked hemp. A unicorn fiber phenotype is a perfect way to capitalize on Dolly's "different" advice.
5. Trevor Hennings, "Male vs. Female Cannabis: How to Determine the Sex of Your Plant," *Leafly*, September 19, 2017, https://www.leafly.com/news/growing/sexing-marijuana-plants.
6. You can also pollinate specific females in the field with the pollen you've gathered, but you'd better tag the plants so you remember which seeds to save.
7. Grignon's own tribe's debut hemp crop got destroyed by the feds in a dubious 2015 raid during the final poisonous puffs of prohibition.
8. "New Mexico," Best States for Business 2018, *Forbes*, https://www.forbes.com/places/nm/.

CHAPTER NINE: COMBINING KILLED THE RADIO STAR

1. Without berry-picking breaks, I'm not sure there would be agriculture. I just don't think it would have developed as a technology. Blackberry in Oregon, huckleberry in Washington, mulberry (plus papaya and such) in Hawaii, raspberry in Vermont. Absent these, we humans would just have kept wandering.
2. Big thanks to Dean Norton at Mount Vernon for directing me to President Washington's journal archives, *The Papers of George Washington Digital Edition* (Charlottesville: University of Virginia Press, Rotunda, 2008). When I searched

Washington's diaries for hemp, I couldn't help noticing that on September 4 in 1765, President Washington had the same plan B issue I did in 2018: "Began to pull the seed hemp," he wrote, "but it was not sufficiently ripe." Too bad he didn't know Edgar.

3. Anne and Eric Nordell, "Cultivating Questions: The Cost of Working Horses," *Small Farmer's Journal* (website), summer 2012, https://smallfarmersjournal.com /the-cost-of-working-horses/.

4. The Salt Creek combine is like my laptop: sturdy, functional, and always on call. It cost Aaron and crew $20,000, rather than $500,000 new. The keep-it-working mind-set is seeing a worldwide comeback in the Right to Repair movement, which aims to take back the option of maintaining our devices and vehicles from the sole purview of the manufacturer.

5. There is already at least one hemp-focused farm-labor-crew management entity in operation in the United States. But that mode is not exactly cutting everyone in on the profits.

6. Fourtenout suggests the folks at EVA (http://evamerica.com) as one source for conversion kits and Farm Hack (https://farmhack.org/tools) as a general informa-tional resource.

7. If our first New Mexico crop goes well, my sweetheart, an amazing seamstress, may even weave some clothing from the bast fiber. Self-sufficiency is the name of the game for this neo–Rugged Individualist family.

8. I really don't like the prevailing industry word for this part of the harvest, *biomass*, because it doesn't convey that we're talking about a precious, sacred, hard-working plant.

CHAPTER TEN: HOW WOULD THE SHAMAN BOTTLE HEMP?

1. That was the nature of our official partnership: We grew the Vermont crop together, then divided it up to be utilized in our respective enterprises. For me in 2019, that included a new pair of Hemp in Hemp retail partners back in the Southwest.

2. Victory Hemp Foods is killin' it with organic hempseed processing on the East Coast, with plans to expand. I love the way its founder Chad Rosen is scaling up from craft to commodity level: He's doing it with regenerative values, farmers in mind, and with levelheaded research into where the seed and seed oil markets are going.

3. Since Hemp in Hemp, for now, is labeled "for external use," taste only matters for my family members and the Gigueres, for our personal use. But I'm honing processes and taste profiles to be prepared to move to an edible product.

4. Many folks, including myself, believe that various cannabinoids might have nutritive benefits in their acid form as well.

5. Mark June-Wells, "Your Guide to Ethanol Extraction," *Cannabis Business Times*, July 11, 2018, https://www.cannabisbusinesstimes.com/article/your-guide-to -ethanol-extraction/.

6. Tim Alchimia, "Complete Guide to Solventless/Non-solvent Cannabis Concentrates," *Alchimia* (blog), April 29, 2019, https://www.alchimiaweb.com/blogen/guide-solventless-non-solvent-cannabis-concentrates/.

7. Alexander Beadle, "Supercritical CO2 Extraction May Cause Loss of Some Terpenes," *Analytical Cannabis*, July 26, 2018, https://www.analyticalcannabis.com/articles/supercritical-co2-extraction-may-cause-loss-of-some-terpenes-306820.

8. As a fellow who likes to drive on vegetable oil power, I've always known these as restaurant cubies.

9. This was a tough decision. It was the first time my product had anything in it that I didn't grow. But to the extent that there might be a benefit to having seed oil and flower from the same plant in the product, I stayed true to Hemp in Hemp's farm-to-table directive: Every bottle contained Samurai flower and some Samurai seed oil. It helped immensely that I knew and trusted Chad: I could rely on the fact that the balance of the hempseed oil in the product was indeed domestically grown and organic. That felt sufficient as a plan B.

10. George Mouratidis, "Beyond CBD: Exploring the Health Benefits of CBN in Cannabis," *Analytical Cannabis*, February 12, 2019, https://www.analyticalcannabis.com/articles/beyond-cbd-exploring-the-health-benefits-of-cbn-311488.

11. As with being in the field during the season, I think some processing experience is important for any putative hemp entrepreneur: I believe in the "know how to do every job in your enterprise" truism. I'm not the best at most of the hemp steps, but I can perform all of them. Except maybe drive a combine.

CHAPTER ELEVEN: HEMP KEEPS YOU THIN

1. Small and Cronquist, "A Practical and Natural Taxonomy," 405–35 (see chap. 3, n. 1).

2. I got this figure by averaging the three test plots discussed in this study, plus what its authors list as the amount of carbon sequestered in a no-till field: https://bit.ly/2lUSlHT.

3. Lori Ioannou, "This Is a $15 Trillion Opportunity for Farmers to Fight Climate Change," CNBC, June 12, 2019, https://www.cnbc.com/2019/06/11/this-is-a-15-trillion-opportunity-for-farmers-to-fight-climate-change.html.

4. We have history's backing with this project. As we discussed in *Hemp Bound*, in many places hemp never stopped playing this prominent nutritive role. In the absence of Wheat Thins, "Roasted whole hempseeds have always been the go-to soccer practice snack in Iran," said Farhoud Delijani, who was a hemp building researcher at the University of Manitoba when I interviewed him. "As a kid in Tehran I didn't know about fatty acids. It was just an everyday after-school treat." The Farsi name for hemp is *Shaah-daaneh*, or "king seed." The original working title for *Hemp Bound*, in fact, was *King Seed*.

5. He is an editor of the former.

6. Laurentiu Mihai Palade et al., "Effect of Dietary Hemp Seed on Oxidative Status in Sows During Late Gestation and Lactation and Their Offspring," *Animals* 9, no. 4 (April 2019): 194, https://doi.org/10.3390/ani9040194.

7. Hemp food for humans recently won GRAS certification. Hopefully the nonhuman side will be figured out not long after the first edition of this book is published.

CHAPTER TWELVE: REBIRTH OF THE BIOMATERIALS ERA

1. I learned this from a fascinating exhibit at the Hemp Museum in Amsterdam.
2. Thank you to the folks at Aleph Objects in Loveland, Colorado, for printing that first hemp goat.
3. Monsanto might seem to be an ag company, but it is a chemical company. Its ads in 1980s *National Geographics* are for stain-resistant carpet coating.
4. Great work by the group Industrial Hemp Illinois in digging up these letters, which you can read at http://antiquecannabisbook.com/chap04/Illinois/IL_IHSherwin Williams.htm.
5. Anslinger's real-life grandniece, Mary, is a cannabis activist who works in the industry.
6. Mitlin is now at Clarkson University, where he and his partners were awarded a $225,000 grant to continue his supercapacitor research in 2018.
7. A lucid, detailed explanation for how supercapacitors work by Chris Woodford is on the Explain That Stuff website at: https://www.explainthatstuff.com/how -supercapacitors-work.html.
8. Not to mention political gamesmanship—on May 29, 2019, Chinese authorities sent stocks tumbling by threatening to cut off exports of 17 rare earth minerals, including several I bet you didn't know are in your everyday life: neodymium and praseodymium.
9. Look up the epoxy precursor epichlorohydrin if you want to be scared.
10. Jon Hamilton, "Study: Most Plastics Leach Hormone-Like Chemicals," NPR, March 2, 2011, 4:07 p.m. ET, https://www.npr.org/2011/03/02/134196209 /study-most-plastics-leach-hormone-like-chemicals.
11. "Japan May Use E-Waste for 2020 Medals," BBC, August 23, 2016, "Business," https://www.bbc.com/news/business-37163074; Tim Hornyak, "Recycling Electronic Waste in Japan: Better Late Than Never," CNN, September 17, 2010, http://travel.cnn.com/tokyo/shop/urban-mining-finding-value-amongst -old-electronics-464333/.
12. As the *cell* in the word *cellophane* alerts us, actual cellophane film, like rayon, is a material that comes from treating various substances, some already plant based, with chemicals such as sodium hydroxide. The plastic film that today wraps, say, your kid's puzzle box, is not generally derived from a regenerative process.
13. Nativia's name sounds similar, but it's an entirely different company from the Nutiva organic hemp food company.

CHAPTER THIRTEEN: FARMER FIBER COLLABORATION

1. Deborah M. Gordon, "What Do Ants Know That We Don't?" *Wired*, July 6, 2013, https://www.wired.com/2013/07/what-ants-yes-know-that-we-dont-the -future-of-networking/.
2. Sunstrand ran into financial difficulties later that season and had trouble fulfilling some of those payments.

3. *Composite* is a generic term for combined industrial components. It can mean plastic or other natural fiber combinations. Flaherty explained the nuances of these terms: "When people claim that something is hemp plastic, they are actually referring to a bio-composite, a compound, that has hemp as the bio-based . . . filler fiber. The polymer [or actual plastic] may be bio-based as well. As in the polymer may have been derived from plant sources, but it requires chemicals and energy to convert the raw material into a polymer. That being said, there is some work being done to produce hemp-derived polymers."

CHAPTER FOURTEEN: CLEANING UP WITH PLANTS

1. Madeline Rubenstein, "Emissions from the Cement Industry," *State of the Planet* (blog), Earth Institute, Columbia University, May 9, 2012, https://blogs.ei.columbia.edu/2012/05/09/emissions-from-the-cement-industry/; "Polluting Our Soils Is Polluting Our Future," Food and Soil Organization of the United Nations, February 5, 2018, http://www.fao.org/fao-stories/article/en/c/1126974/.
2. A company in Kentucky called Fibonacci is already working on a "Hemp Wood."
3. A tip: A small amount of vitamin C added to the mixture (maybe 1 percent) can help combat both mildew and termites.
4. A note on sourcing your hempcrete binder: Most hemp builders use hydrolyzed lime. Some will want to sell you proprietary binding mixes. But limestone isn't found everywhere, so we see companies shipping the rock between continents to customers. I'm lucky—we have great regional lime sources in the Southwest. In Hawaii, a volcanic cinder binder company has real potential for supplying the coming biomaterials construction era in the Aloha State. Bottom line, research your binder the way you do all your inputs, and try to source it as regeneratively as possible.
5. GlobeNewswire, "Environmental Remediation Market Size to Reach USD 122.80 Billion by 2022," news release, January 17, 2018, https://www.globenewswire.com/news-release/2018/01/17/1295780/0/en/Environmental-Remediation-Market-Size-to-Reach-USD-122-80-Billion-by-2022-Zion-Market-Research.html.
6. Steve DeAngelo, cofounder of Harborside dispensaries, heard me talking about soil building as an industry, and directed me to this proposed venture in that area, which aspires to pay farmers $15 per metric ton of carbon sequestered: https://www.cnbc.com/2019/06/11/this-is-a-15-trillion-opportunity-for-farmers-to-fight-climate-change.html.
7. Sonia Campbell et al., "Remediation of Benzo[a]pyrene and Chrysene-Contaminated Soil with Industrial Hemp (*Cannabis sativa*)," *International Journal of Phytoremediation* 4, no. 2 (2002): 157–68, https://doi.org/10.1080/15226510208500080.
8. Jenny Hopkinson, "Can American Soil Be Brought Back to Life?" *Politico*, September 13, 2017, https://www.politico.com/agenda/story/2017/09/13/soil-health-agriculture-trend-usda-000513.
9. "Report Sounds Alarm on Soil Pollution," Food and Agriculture Organization of the United Nations, May 2, 2018, http://www.fao.org/news/story/en/item/1126971/icode/.

CHAPTER FIFTEEN: HERDING RUGGED INDIVIDUALISTS

1. Gar Alperovitz, "The Cooperative Economy," *Orion*, May 21, 2014, https://orion magazine.org/article/the-cooperative-economy/.
2. "What Is a Cooperative?" International Cooperative Alliance (website), https://www .ica.coop/en/cooperatives/what-is-a-cooperative.
3. U.S. Small Business Administration, *Frequently Asked Questions About Small Business*, August 2018, https://www.sba.gov/sites/default/files/advocacy/Frequently-Asked -Questions-Small-Business-2018.pdf.
4. Like everyone in the hemp industry, the FPS is waiting to see how and when USDA and FDA hemp guidelines unfold, which could affect their phase two schedule.
5. Ernest Small and David Marcus, "Hemp: A New Crop with New Uses for North America," in *Trends in New Crops and New Uses*, eds. Jules Janick and Anna Whipkey (Alexandria, VA: ASHS Press), 284–326.
6. *Carbon miles* means the distance a product travels to market when powered by fossil fuels. Most American food today has 1,500 carbon miles embedded in it from farm to table.
7. Because we hand-harvested in 2018. Earlier we harvested with a biofuel-powered combine.

CHAPTER SIXTEEN: GREEN CHILE HEMP

1. You can also answer, "Christmas tree," which, as you might have guessed, is both red and green.
2. "Nutrition and Health," New Mexico Green Chile Company (website), 2007, https:// greenchileco.com/nutrition-health/; "Supplement Guide: Capsaicin," Living with Arthritis, Arthritis Foundation, https://www.arthritis.org/living-with-arthritis /treatments/natural/supplements-herbs/guide/capsaicin.php.
3. Some vintners prefer the term *appellation* for the concept we're discussing.
4. Brad Evans, "Vermont Named Craft Beer Capital of U.S.," NBC5, January 3, 2019, https://www.mynbc5.com/article/vermont-named-craft-beer-capital-of-us/25663312#.
5. Mike Snider, "Craft Beer Sales Continue Growth, Now Amount to 24% of Total $114-billion U.S. Beer Market," *USA Today*, April 2, 2019, https://www.usatoday .com/story/money/business/2019/04/02/beer-sales-stay-flat-craft-beer-grows -share-114-b-us-market/3341312002/.
6. The Bradfords also grow okra and are branching into hemp.
7. Phil James, "The Man Behind the Richmond Company That Won't Stop Growing," *Richmond Confidential*, November 23, 2014, https://richmondconfidential.org/2014/11 /23/the-man-behind-the-richmond-company-that-wont-stop-growing-2/.

CHAPTER SEVENTEEN: THE FRIENDLY FUNGUS AND THE HAIRNET

1. Christina Rice et al., *Cottage Food Laws in the United States* (Cambridge, MA: Food Law and Policy Clinic, 2018).
2. Stephen Harrod Buhner, *Sacred and Herbal Healing Beers: The Secrets of Ancient Fermentation* (Boulder, CO: Brewers Publications, 1998), 63.

3. Rebecca Kreston, "This Ain't Yo Momma's Muktuk: Fermented Seal Flipper, Botulism, Being Cold and Other Joys of Arctic Living," *Body Horrors* (blog), *Discover*, August 14, 2011, http://blogs.discovermagazine.com/bodyhorrors/2011/08/14 /this-aint-yo-mommas-muktuk-or-fermented-seal-flipper-botulism-being-cold -other-joys-of-artic-living/#.XcLhW5JKjY0.

4. Buhner, *Sacred and Herbal Healing Beers*, 63.

5. Jereme Zimmerman, *Make Mead Like a Viking: Traditional Techniques for Brewing Natural, Wild-Fermented, Honey-Based Wines and Beers* (White River Junction, VT: Chelsea Green Publishing, 2015), 69.

6. Lindsey Hoshaw, "Nine Things You Didn't Know About the Fall Grape Harvest," *Bay Area Bites*, KQED, October 11, 2013, https://www.kqed.org/bayareabites/71880 /nine-things-you-didnt-know-about-the-fall-grape-harvest.

7. Michael Pollan, "Some of My Best Friends Are Germs," *New York Times Magazine*, May 15, 2013, https://www.nytimes.com/2013/05/19/magazine/say-hello-to-the -100-trillion-bacteria-that-make-up-your-microbiome.html.

8. Bruce Schneier, "Beyond Security Theatre," *New Internationalist*, November 1, 2009, https://newint.org/features/2009/11/01/security.

9. "Food-Borne Illnesses," *Frontline*, April 18, 2002. Elisabeth A. Mungai, Casey Barton Behravesh, and L. Hannah Gould, "Increased Outbreaks Associated with Nonpasteurized Milk, United States, 2007–2012," *Emerging Infectious Diseases* 21, no. 1 (January 2015), https://doi.org/10.3201/eid2101.140447.

10. Zhang Jianchun, *Natural Fibers in China* (Rome: IYNF 2009 Symposium, 2008).

11. There was an important bit of good news in Gottlieb's note, in that it codified what humans already have known for 12 millennia—that hemp is officially Generally Regarded as Safe (GRAS).

12. "Gottlieb to Talk to Congress Regarding Approaches to CBD," *Whole Foods Magazine*, February 28, 2019, https://wholefoodsmagazine.com/news/main-news /gottlieb-to-talk-to-congress-regarding-approaches-to-cbd/.

EPILOGUE: THE REGENERATIVE ENTREPRENEUR

1. "Industrial Hemp Markets to Reach USD 13.03 Billion by 2026," "Reports and Data," *Globe Newswire*, May 6, 2019, https://www.globenewswire.com/news -release/2019/05/06/1817648/0/en/Industrial-Hemp-Market-To-Reach-USD -13-03-Billion-By-2026-Reports-And-Data.html.

2. Russell T. Graham and Theresa B. Jain, *Ponderosa Pine Ecosystems* (Albany, CA: Pacific Southwest Research Station, USDA Forest Service, 2005), https://www.fs.usda .gov/treesearch/pubs/27254.

3. "Global Land-Ocean Temperature Index in 0.01 degrees Celsius: 1951–1980," NASA document, July 2019, https://data.giss.nasa.gov/gistemp/tabledata_v3/GLB.Ts+dSST.txt.

4. Olivia Rosane, "CO2 Levels Top 415 PPM for First Time in Human History," *EcoWatch*, May 13, 2019, "Climate Change," https://www.ecowatch.com/co2 -levels-top-415-ppm-2637007719.html.

5. Another sign of hemp's return to normalcy is the evolution of state program signs and permits themselves. These reflect the way state program coordinators ask

farmers to announce their official status. In three years, it's gone from weird fencing rules and crazy amount of signage (as though announcing a superfund site) to my New Mexico permit this year, which is a piece of paper containing my name and our Department of Agriculture's seal. I can mount it outside or not as I like. I'm actually using it as a mouse pad at the moment.

INDEX

Note: Page numbers in *italics* refer to figures.

AMERICAN HEMP FARMER

cooperative model (*continued*)
  types of co-ops, 223
  See also Fat Pig Society workers co-op
copper steam distillation, 172–73
corn acreage, 137–38
costs
  clones, 81
  cold ethanol processing, 86, 172
  combines, 288ch9n4
  cultivation per acre, 84
  fiber-processing equipment and facilities, 210
  flower processing, 169–170
  genetics development, 82
  Hemp in Hemp, 166–67
  nano sheets of hemp fiber, 199
  pack animal vs. combine harvesting, 146
  seed, 81–83
  seed-processing equipment, 87–88
  soil building, 38–40
  THC testing, 126–27
cottage food law, 256, 257, 258
cover crops, 38, 42, 117
craft model
  branding and marketing considerations, 245–48
  defining sector by annual production volume, 256
  parameters for regenerative distribution, 250–54
  regulatory considerations, 262–64
  sustainable production limits, 252
  wholesale markets vs., 17–18
  See also value-added products
Cranshaw, Whitney, 105
Crew, Shaun, 24
Cronquist, Arthur, 55–56, 57–58, 74
crop contamination, during planting, 97
crop insurance, 130
crude
  cold ethanol processing of, 170–72
  definition of, 68
  potential revenues, 226, 235, 236–37
  toll processing of, 85–86
cultivation of hemp
  clones, 77–79
  learning your own field, 112–13
  midseason concerns for Colville 2017 crop, 114–122
  outdoor vs. greenhouse cultivation, 81
  permits for, 59

plant growth rates, 109
polyculture approach, 108
regenerative agriculture benefits, 11–12, 19, 183–84
transplanting of plugs, 78
tri-crop approach, 77, 81, 83
watering protocols, 52, 110–11
weeding protocols, 107–10
  See also acreage of hemp under production; dioecious crops; planting; sinsemilla (female-only) crops; soil building
curing of harvested flowers, 156

DeAngelo, Steve, 6–7, 291n6
decarboxylation of hemp flowers
  advantages and disadvantages of, 173–74
  benefits of, 168
  bubbling pattern, 177
  Fat Pig Society, 53
  processing of 2018 Vermont crop, 176–79
  steps in, 169–170
decortication, 89, 209, 210–11
Delijani, Farhoud, 289n4
Devane, Bill, 287ch8n3
Dickey-John Grain Analysis Computer, 124–25
diet. *See* health and diet aspects of hemp
digestion, microbial balance and, 188–89
digital marketing, 249
dioecious crops
  pollen gathering from, 127–29
  potential revenues, 87
  sinsemilla cultivars vs., 17, 80
  THC testing policies, 69–70, 285n8
direct marketing approaches, 249–250
distribution platforms
  parameters for regenerative distribution, 250–54
  small batch processing needs, 15–18, 249–250
Doherty, Ryan, 210
Dr. Bronner's, 253–54, 286ch6n1
drying processes
  preparation for, 123–25
  silo drying of crop, 125, 149, 155, 287ch8n1
Dyck, Grant, 151

Eden Labs, 86, 171
edible hemp products. *See specific products*
electric-powered equipment, 152–53, 157, 200
electronic devices, pollution from, 204
endocannabinoid system, 6

— 298 —

ABOUT THE AUTHOR

Amanda Gorski

Doug Fine is a solar-powered goat herder, comedic investigative jour-
nalist, and pioneer voice in cannabis/hemp and regenerative farming. He
has cultivated hemp in four US states, and his genetics are in five more.
He's an award-winning culture and climate correspondent for NPR, the
New York Times, and the *Washington Post*, among others. His previous
books include *Hemp Bound*, *Too High to Fail*, *Farewell, My Subaru* (a
Boston Globe bestseller), *Not Really an Alaskan Mountain Man*, and *First
Legal Harvest*, a monograph that was printed on hemp paper. His print
and radio work, United Nations testimony, and TED Talk can be found
at dougfine.com, and his social media handle is @organiccowboy.